INFORMAL SCIENCE LEARNING: WHAT THE RESEARCH SAYS ABOUT TELEVISION, SCIENCE MUSEUMS, AND COMMUNITY-BASED PROJECTS

Valerie Crane
Milton Chen
Stephen Bitgood
Beverly Serrell
Don Thompson
Heather Nicholson
Faedra Weiss
Patricia Campbell

For information about this book, write or call:
Research Communications Ltd.
990 Washington Street
Suite 105
Dedham, MA 02026
(617) 461-1818

This material is based on work supported by the National Science Foundation under Grant No. ESI-9254384. Any opinions, findings, and conclusions or recommendations expressed in this material are those of the authors and do not necessarily reflect the views of the National Science Foundation.

Printed in the United States of America

FOREWORD

I would like to thank all of the dedicated people who contributed to this book. First, to George Tressel for his tireless reviews of each chapter and important contributions in focus and reference. Alan Friedman, Director of the New York Hall of Science, also gave helpful suggestions on references and editing of the introductory and final chapters of the book.

I am also grateful to those individuals who provided reviews of the chapters and/or participated in interviews for the final chapter of the book. This distinguished list of individuals includes Jerry Wheeler of AAAS, Judy Diamond of the University of Nebraska, Michael Templeton, Media Consultant, John Howard Falk, Museum Consultant, Chan Screven of the University of Wisconsin-Milwaukee, Jerry Bell of AAAS, Bruce Lewenstein of Cornell University, June Foster of TERC, Alan Friedman of the New York Hall of Science, Gida Wilder of Children's Television Workshop, Ron Slaby of Harvard University, Keith Mielke of Children's Television Workshop, Lynn Dierking, Consultant, Don Agostino of Indiana University, Elsa Feher of R.H. Fleet Space Theatre and Science Center, Harris Shettel, Museum Consultant, Jon Miller of Chicago Academy of Sciences, Bonnie Van Dorn of ASTC, Richard Lerner of Michigan State University, Mark St. John of Inverness Research Associates, Minda Borun of the Franklin Institute, Shirley Malcolm of AAAS, and Laura Martin of Children's Television Workshop.

And, finally, thanks to my colleagues Tom Birk for his assistance with the mailing and preparation of the annotated bibliography and Susan Young Domolky for her editing of the final manuscript.

CONTENTS

PREFACE

It is easy to forget how recently the concept of informal learning has developed. We have always acknowledged the importance of parents and early experience, and we have long recognized that the best "school" is a *student and teacher on opposite ends of a log.*

Few of us would deny the effects of our parents' influence . . . or our hobbies . . . or early experiences of travel and role models. We have long felt that participation in Girl Scouts and Boy Scouts is an important part of growing up, and many of us can fondly recall the excitement of a *secret decoder ring.* We can still remember a *particular* day at the zoo or a visit to see DINOSAURS in the museum! But we don't remember *learning* something: *learning* is that painful thing that happens in school.

Efforts to provide such experiences and emotional involvement systematically, and to integrate them with other learning, are comparatively new—as are the tools and institutions to do so. But we are increasingly aware that most of the time, most of us learn most of what we know, outside of school; and success in schooling usually reflects a combination of nature and nurture, not simple "brains and talent." It is nice to be bright and talented, but prize-winning flowers do not usually grow in the shade of an urban alley.

A growing body of behavioral research continues to emphasize the importance of early and continued informal learning—from the earliest parental influences; to the development of naive concepts about nature; to the development of strategies for learning and problem solving; to lifelong patterns of self confidence and intellectual curiosity.

While the *practitioners* of informal education have been developing a vocabulary of new tools and techniques for teaching through museums, broadcasting and hands-on activities; social and behavioral psychologists have been building a theoretical framework that helps to explain the importance and *role* of such experience. Unfortunately, professionals in the two fields do not often talk to each other.

Learning is *not* a sequential process of pouring information into a mind; it is not analogous to programming a computer. Rather, the human mind depends upon storing, and calling upon, a vast network of parallel and intricately related bits of information. We have an ability to store experiences from many occasions, many sources, and many times . . . and when they are needed, to recall and relate *just those items* that are useful to solve a problem. How do we do this?

Recent research has shown that when random groups of letters are flashed on a screen, the visual area of our brain "comes to life"; but if the letters form a word like "boat" or "accident," *many parts* of the brain begin to become active . . . presumably they are making links and helping us to *use* the information. Unconsciously throughout our lives, we are busy developing schema, collecting and linking networks of related information together with patterns of using that information when something is important to us.

The ability to probe and manipulate these schema—to draw on distant and superficially unrelated experiences, to retrieve arcane information, and then to apply all this to complex abstract problems—is uniquely human. Arguably, our most important learning concerns building these schema of experience and developing the facility and self confidence for application. Yet we are only beginning to understand the process and there is too little dialogue between theoreticians, who are developing this model, and the practitioners who produce television programs, make exhibits, and work directly with children after school.

As a result, it is not surprising that an overview of research across a very new and burgeoning field tends to document the study of *techniques, effectiveness and short-term impact* rather than the more difficult and subtle issues of thinking and problem solving. At present, we tend to measure changes in knowledge, attitude and skill at the individual level, and then to relate this to the inarguable scale and cost-effectiveness of mass communication techniques. In effect, *individual excitement and success* times the *number of individuals* equals *impact and cost effectiveness.* In time, we can expect to integrate such pragmatic studies with an increasingly sophisticated model of the learning process, both improving the techniques and documenting the critical role of informal learning to the success of all education.

For most purposes, the practice of informal education as we know it spans only about two generations. Prior to the turn of the century, for example, the concept of a museum was greatly different. First as "Wonder rooms" where wealthy aristocrats entertained their guests with incongruous collections of such "Gee whiz!" oddities as skulls, stuffed animals and gallstones; then as arcane academic study-collections; then as public displays of taxonomic enterprise . . . museums were a place to see, not learn. With the advent of the Deutsches Museum, however, and later the development of hands-on science teaching in the 1960s, the purpose of museums changed radically. Today, no one questions *whether* a museum should teach, the question is *how?*

The same is true of broadcasting. In the brief time since television appeared we have proven its immense power to reach and teach millions of children, to entice complex abstract thinking, and to engender significant behavioral and intellectual change. We are well into the second generations of *Sesame Street, Reading Rainbow, 3-2-1 Contact,* and *Square One TV* so it is no surprise that our techniques of informal learning are envied and emulated by other countries.

Youth groups have always relied upon hands-on learning, but they are beginning to see their role as more important than simple recreation. In a world where parent attention is ever more difficult, the role of informal learning in organizations like Girls Inc. can make a critical difference in a child's social, personal and intellectual development. The girl who considers math and science "child's play" is ready to take on the competitive and elitist world of science and research.

Most of the research described in this state-of-the-art summary concerns technique—audiences, understanding, presentation and effectiveness—this is the state of the field. By making this so clear, the discussion should help to focus greater attention on the urgent need for links to a theoretical base; models and assessment of what we are *trying* to do, what we *expect* to happen, and the critical impact on success in other learning.

If we can continue to establish such links . . . between research on techniques and research on learning processes and goals, informal learning will become recognized as a critical component of successful education.

ABOUT THE AUTHORS

Stephen C. Bitgood holds a Ph.D. from the University of Iowa and is Professor of Psychology at Jacksonville State University and President of the Center for Social Design. He is editor of *Visitor Behavior* and is on the editorial boards of *Environment and Behavior* and *ILVS Review: A Journal of Visitor Behavior*. In addition to working with a number of museums and zoos on exhibition evaluation projects, he is involved in research projects designed to better understand the visitor experience.

Patricia B. Campbell, Ph.D. has directed Campbell-Kibler Associates, an educational research and evaluation consulting firm for the past thirteen years. Formerly an associate professor of research, measurement and statistics at Georgia State University, Dr. Campbell's research and evaluation interests center around issues of gender and ethnicity in math and science in formal and informal settings. She has worked with such groups as ASTC, Girls Incorporated, the Children's Museum (Boston) and the Museum of Science (Boston).

Milton Chen, Ph.D., directs the KQED Center for Education and Lifelong Learning (CELL), established in 1991. The San Francisco-based Center utilizes public television programming and services to extend educational opportunities for learners of all ages. Dr. Chen joined KQED as Director of Instructional Television (ITV) in 1987.

For the past 20 years, Dr. Chen has worked in program development, audience research, and community outreach for public television's most prominent children's series including *Sesame Street, The Electric Company, Square One TV,* and *Ghostwriter,* all produced by the Children's Television Workshop in New York. From 1978 to 1980, he served as Director of Research for CTW's *3-2-1 Contact* on science and technology for 8- to 12-year-olds. From 1985 to 1987, Dr. Chen was an assistant professor at the Harvard Graduate School of Education in Cambridge, teaching and conducting research on educational media.

He currently serves on advisory committees for the Council of Chief State School Officers, the Office of Technology Assessment of the U.S. Congress, Children's Television Workshop, PBS, and Scholastic Productions. In October, 1993, he traveled to South Africa to work with educational media groups preparing for the transition to a democratic society. He is also helping to develop a Sesame Street theme park concept. Dr. Chen is a member of the Board of Directors of the San Francisco Bay Girl Scout Council and has served on the Board of ARC

Associates, an educational research firm specializing in the needs of language-minority students.

Valerie Crane, Ph.D., is President of Research Communications Ltd., a communications research company in Dedham, Massachusetts which she founded in 1980. Research Communications Ltd. conducts research on a wide range of educational media projects, public television programs, broadcast and cable television programs and networks, international media projects, and science museums. She currently serves on the Committee for Public Understanding of Science and Technology for the American Association for the Advancement of Science.

Heather Johnston Nicholson, Ph.D. is Director of the National Resource Center of Girls Incorporated, a national youth and advocacy organization in Indianapolis that enables girls to become strong, smart and bold. Since 1985 she has been the principal investigator of the research about girls in math and science undertaken as part of the development of Operation SMART (**S**cience, **M**ath **a**nd **R**elevant **T**echnology), a major initiative of Girls Incorporated to sustain girls' interest and participation in these fields. Dr. Nicholson also has been principal investigator for the longitudinal study of the effectiveness of the Girls Incorporated Preventing Adolescent Pregnancy Project and she has taken a major role in the design and evaluation of programs on substance abuse prevention, HIV education and health promotion, as well as in the organization's overall approach to gender equity, program development, and program evaluation.

Beverly Serrell holds an MA in science teaching in informal settings and a BS in biology. She has been an exhibit and evaluation consultant for the last 14 years with art, history, natural history and science museums, as well as zoos and aquariums. In addition, she was formerly a museum education department head for eight years, and had shorter stints as a high school science teacher and a research lab technician. She has been a frequent museum visitor all her life.

Don Thompson holds a Ph.D. from the University of Wisconsin-Milwaukee and is currently a research associate with the Center for Social Design. He is a contributing editor of *Visitor Behavior* and an editor of the collected papers of the annual Visitor Studies Conferences. He is involved in visitor research projects at several museums.

Faedra Lazar Weiss is Research Associate at the Girls Incorporated National Resource Center, where she is involved in research on the issues that girls and young women face in growing up confident and competent. As a member of the Girls Incorporated Needs Assessment and Evaluation Team, she provides training and technical assistance to affiliates in designing and implementing program evaluation. She is currently conducting a telephone survey of Girls Incorporated staff throughout the country, evaluating their replication of Operation SMART (**S**cience, **M**ath and **R**elevant Technology), the organization's national initiative encouraging girls' interest and participation in these important fields.

CHAPTER

An Introduction to Informal Science Learning and Research

Valerie Crane, Ph.D.

Early in September, 8-year-old Cindy rises and gets ready for school. While waiting for her mother, she goes into the living room and turns on the TV, which shows the wonders of the migration of whales. Cindy is intrigued with the size and power of the whales and she reflects on this as she heads for school. Months later Cindy recalls this moment when her teacher assigns the class to small groups which sort and classify the different foods that whales eat. As part of this unit, her assignment is to write an ecology essay on the food chain. Over the Christmas holidays, her uncle talks about a parent training workshop, "Turn Your Kids on to Science." He spends some time with her and shares a map on the migration patterns of whales. Later that spring, Cindy's mother sees a story on the local news about some beached whales about a half-hour from where they live. She calls Cindy in to watch and they talk about how whales are the biggest mammals and how they would all die if it weren't for the littlest microscopic animals in the sea that they eat. Cindy writes the local television station to find out more. The station sends her a brochure on whales, explaining the phenomenon. The brochure suggests a visit to the aquarium to look at a new exhibit which shows a baby dolphin being born. She asks her mother if they can visit the aquarium some time. Cindy's mother remembers her request several months later when they are planning activities for a rainy day during the summer. Cindy notices how the exhibit explains how big fish eat smaller fish—something she had written about in her ecology paper for school.

This scenario involves a range of rich learning experiences for Cindy. It is, no doubt, an ideal world where parents are involved in their child's education, and are willing and able to make the linkages between everyday informal learning experiences and school. In Cindy's life, there are the school-based experiences that come under the purview of formal science learning. And there are activities that take place outside school—informal science learning activities that support and expand on what she is learning in school. As is usually the case with informal learning, the link to the formal learning experience was neither planned nor accounted for.

While the informal learning experiences for Cindy seem especially rich, they mirror the way in which most children experience and learn about the world in which they live. Children naturally make the connections between what they are learning from many different sources. They build hypotheses about how the world works long before they go to school, and continue that learning process throughout their lives. Support for these connections exists in research on informal learning. For example, Research Communications Ltd. (1987) conducted a study of *3-2-1 Contact* which showed that six in ten of the parents in the study were able to provide specific examples of how their children made links between what they had seen on *3-2-1 Contact* to other events in their lives such as a museum exhibit they saw or book they read at a later date.

When we speak of learning, one assumes the focus is on school-based learning. School is on the schedule and in the budget. Much attention has been paid to how the school environment fares in providing our children with science education. The results have not been encouraging, although there are many examples of extraordinary teaching and learning going on throughout this country (Suter, 1993). Governmental funding agencies (e.g., the Department of Education and the National Science Foundation) have funded a substantial number of projects devoted to improving students' performance in school science. Although it is difficult to estimate the amount of support available for informal learning, it is well known that much more limited support is available to create informal science learning experiences and even less, if any, support is available to understand its contribution to the education of both our youth and adult populations.

After Cindy's school experience of classifying the food of the whale, her teacher gave her a "test" which showed her mastery of what she had learned. Comfortable that Cindy had achieved the lesson's objective, her teacher moved her on to another lesson. But no one ever bothered to evaluate the contribution of her informal learning experiences, including how these experiences set the stage for other learning. A number of unanswered questions arise when we look more carefully at what Cindy has experienced outside school.

- *What experiences piqued Cindy's interest in whales?*
- *What contribution did the children's TV program on whales make to Cindy's understanding of whale migration?*
- *What was the impact of the map activity she engaged in with her uncle?*
- *What has the family been learning about science by watching science programs on television?*
- *How did Cindy and her mother react when they saw the news story on the whales?*
- *How did the visit to the aquarium reinforce her earlier learning?*

These questions provide a framework for what we need to learn about these informal learning experiences. What is really going on with informal learning? What do we need to know about it?

That is what this book is about: Initiating the inquiry into what informal science learning is, what it means to people in their lives, what we understand about it, what we don't know about it, and why we don't know more. Few efforts, if any, have taken a direct look at what we know and do not know about the impact of informal science learning on the public.

For the purposes of this book, **informal science learning** refers to activities that occur outside the school setting, are not developed primarily for school use, are not developed to be part of an ongoing school curriculum, and are characterized by voluntary as opposed to mandatory participation as part of a credited school experience. Informal learning experiences may be structured to meet a stated set of objectives and may influence attitudes, convey information, and/or change behavior. Informal learning activities also may serve as a supplement to formal learning or even be used in schools or by teachers, but their distinguishing characteristic is that they were developed for out-of-school learning in competition with other less challenging uses of time; these uses will be the primary focus of the studies cited in this book.

There are many informal learning media including exhibits and demonstrations in museums, aquariums, and zoos: television, radio, and community-based programs, books, magazines, hobbies, and newspapers. In this book, we will consider three of these domains of informal learning media: television and radio programs; science museums includ-ing technology centers, aquariums, and zoos; and community-based projects.

The Case for Informal Learning

Educators and scientists might read Cindy's story and agree that such learning experiences make an important contribution to her understand-ing of the world around her. Most would agree that we should be working towards creating more of these experiences to enrich children's learning environments. Most would also agree that these kinds of informal learn-ing experiences in science should be available to adults as well. And many would make the case for developing informal learning experiences that include both parents and children, thereby providing a supportive family environment for science learning.

This is the first and most compelling statement that one can make for informal science learning: that science learning should and does occur not only in school but outside as well, not only during the limited years we are in school but throughout our lifespan. In fact, informal science learning addresses a number of key challenges that scientists and science educators struggle with on a regular basis. For example:

1. How do we foster the idea among the public that science is an important endeavor that has a positive impact on our lives?

A study of *Square One TV*, a math service for 8- to 12-year-old children, by Research Communications Ltd. (1989) showed that in homes where other informal math learning experiences were common, children were more likely to watch *Square One TV*, than in homes where informal math learning experiences were unknown. In other words, there seemed to be a predisposition among some families (particularly those with a higher level of education) to pursue informal math learning experiences with their children. The challenge for informal learning practitioners is to provide all families with access to these resources, including those with lower levels of education, with less motivation to engage their children in informal science learning, and with less self-confidence in acting as science learning facilitators for their children. The Family Math Program that started in Berkeley, California has met this challenge in cities across the country and is discussed in Chapter Four.

2. If experiences outside school influence the pursuit of science, how can we leverage these experiences to support the pursuit of science?

Many professionals in the fields of science and mathematics believe that science and mathematics are meaningful and worthwhile endeavors and should figure importantly in the information world of the public. Few of us can escape science and math in our lives even if we work hard to do so. When a family member becomes ill with debilitating back trouble, a heart attack, a stroke, or cancer, what kinds of science information does the family seek or want to know? When a toxic waste dump is reported in your area, what steps do you take to learn more about what this means? How do the informal learning networks help the public understand these science-based issues that they face in their daily lives?

Scientists would argue that creating appreciation for science not only empowers the public to deal with these challenges in their lives, but that the scientific process itself is a meaningful endeavor for all to appreciate. One also could argue that as more positive attitudes develop about science, the more likely we are to see the public taking science issues seriously, the more valued these careers appear to be to our youth of all ethnic and gender groups, and the more likely mainstream media are to include science as part of what they communicate to the public.

3. How can we maximize the flow of talented youth into the sciences as a course of study and career?

Perhaps the most compelling evidence that the schools are failing our children is the repeated finding that girls and minorities are underrepresented in the roll call for science careers. The schools seem to compound this problem rather than alleviate it. Girls and minorities continue to fail at rates greater than their peers and choose not to enter science as a career (see Chapter Four).

Yet, when career decisions were examined more closely, many cited out-of-school or informal science experiences as playing a key role in their decision (Bloom, 1985). This suggests that there are many opportunities to influence the decision-making of youth outside school as they move through their school years. In fact, in Chapter Four, Nicholson, Weiss, and Campbell make the case for community-based projects targeting girls and minorities to pursue science-based careers.

4. How do we continue to reach people with information about science after they leave a formal learning environment?

It is immediately obvious that most individuals will spend far more years as informal learners of science than as formal learners of science. In fact, most individuals will spend on average four times as many years out of school as in school. Learning certainly does not end at the age of 16, 18, or 22. So informal learning becomes an even more important part of the educational landscape as Cindy grows older, especially if she does not choose a career in science.

Even if Cindy goes to college and studies science there, she will forget much of what she learned in school. After college, few of us could pass a college exam on biology, chemistry, or physics even if we studied these in college. Informal science learning offers an opportunity not only to foster learning but to mitigate this loss of knowledge.

5. How do we keep the public updated on what is happening in science?

If science learning ended with formal schooling, then those of us studying science in the 1960s would be oblivious to the endangerment of many species of animals, the erosion of the tropical forest, the deterioration of the ozone layer, and toxic waste problems in local communities around the country. Yet each of these issues has the potential to irrevocably change our lives. What media can or should be used to inform the public of new scientific developments?

6. How do we create an informed public, however small, that will become involved in science issues?

Some science educators and/or researchers believe that resources should be focused on the general public. But Miller (1987) maintains that public education efforts should focus on sustaining science literacy and satisfying the needs of the 20% who will actively seek science information throughout their lives. Whichever group is the target of informal learning, the question remains: "How do we create and sustain an informed public?"

We could answer each of the questions posed above through the systematic development of informal science learning experiences. Informal learning already plays an essential role in educating the public. Through our schools, science education currently underserves important segments of the population (e.g., girls and minorities). The schools have made it clear that they, alone, cannot assume the burden of creating an informed public in science. It is essential that we examine alternative ways of creating public understanding of science, that we do it more intensively and vigorously than we do now, and that we do so with all the standards of excellence and priority that are afforded our formal educational system.

Research on Informal Learning

The structural differences between informal and formal learning affect the kind of audience research and evaluation that is done and the goals for the research efforts. Informal learning programs rely on voluntary participation and compete with a wide array of leisure-time activities. Therefore, the first challenge is to attract and then to hold an audience. This means that participation will depend upon providing a meaningful and/or enjoyable experience. Informal learning projects devote considerable resources to accomplish this, particularly as the distractions increase and as participants pay to participate. One way to assure project success in this regard is through the use of research and evaluation.

Formative Research and Evaluation

The most common form of research performed on informal learning projects is formative research, conducted both before and during their development. Questions about their content design, format, and marketability all need to be addressed in order to ensure that there will be an audience for these informal learning experiences. Formative research and evaluation are relatively common tools in developing television series, museum exhibits, and community projects, all of which face the primary challenge of attracting and holding the attention of their intended audiences.

Formative research often has a significant impact on the project development process itself. Based on experience conducting hundreds of these kinds of studies over the past twenty years, the author has found that it is not uncommon for the scientist, producer, or exhibit developer to revise his or her thinking about what the audience can realistically absorb in a half-hour television program, museum exhibit, or a workshop for parents. So formative research is a useful tool, implemented by practitioners to craft messages and get to know their audiences.

Formative research is conducted before and during the development of the informal learning experience, but it does not determine impact or outcomes once the project is completed. While formative research and evaluation will be discussed to some extent in each of the three chapters, the thrust of this book has been to focus on the impact that informal science learning experiences have on the public.

Summative Research and Evaluation

While formative research and evaluation is conducted during project development, summative research and evaluation is conducted once the project is completed and implemented. Summative evaluation addresses questions about the outcomes and impact of the project. These data are used to determine the effectiveness of learning projects and can also prove useful in providing direction in the development of subsequent projects.

Although impact is considered a key issue for informal science learning, summative evaluation is conducted much less often than formative research for several reasons. First, funding for projects is usually awarded to a producing organization which is faced with developing materials that will attract and hold an audience. When balancing the benefits of research that will assist in effective project design as opposed to measuring results at the end of the project, it is not surprising that practitioners opt for the former. Second, summative research is very costly, and funders have, to date, not made significant funding available on a regular basis for the pursuit of summative research and evaluation. Third, some producing and funding agencies have been reluctant to risk "negative results" from impact research and evaluation. Fourth, setting goals for informal learning projects often sparks controversy. The goals for informal learning projects are often vague, overly ambitious, nonexistent, or ignored once the project is underway. Some consider goal-setting to be a proposal writing activity but not a project development activity.

Among those who feel that the goals are important, there are differing opinions about what the goals of informal learning should be. In fact, when the two dozen or more professionals were interviewed for the final chapter of this book, almost all of those interviewed expressed concerns about the appropriate questions to ask about informal learning. Some felt the primary goal of informal learning should be to inform or to educate. Some focused on cognitive outcomes including knowledge gains or conceptual understanding. Yet others emphasized the importance of changing perceptions and attitudes but noted the difficulties in measuring changes in the affective domain. Finally, some others targeted behavior change as important for informal learning while some felt that this was inappropriate or unrealistic.

Design Challenges in Informal Science Learning Research

Another reason that impact studies are not common in the research literature is that they are difficult and complicated to design and execute. There is no set curriculum that all informal learners are exposed to, making it difficult to work from a uniform outcomes model that characterizes school-based research.

Capturing the Target Audience

Sampling poses special challenges in informal learning research. It is difficult to capture learners physically for research once they leave the museum or community-based program. Finding audiences that have been exposed to specific informal science television programming is even more difficult, especially for a specific science program that might have reached a large number of viewers but a small percentage of the total population.

Yet the integrity of the research data will be determined by the adequacy of the sample itself. In much of social science research, samples are drawn without much consideration of their projectability to targeted audiences. With museum research, random selection of visitors through the museum seems appropriate as long as we understand that the results can only be generalized to those who self-selected to attend the museum. The results could not be generalized to a general population.

With community-based projects, there is again a self-selection bias introduced by the very nature of the programs themselves. For example, what bias is introduced when young people are recruited for these programs? Are they already more motivated than their peers? It is important to determine that prior to generalizing the results of these studies.

Since television needs to recruit samples in order to conduct research, there are more opportunities to establish truly random procedures for selection with the hopes of generalizing the findings to a universe that mirrors the universe of all potential viewers. However, does cooperation to participate in a study bias the sample in some way? Random selection of samples certainly increases chances for an unbiased sample. And while this can be accomplished by setting quotas that mirror the target audience characteristics such as education and ethnicity, it can be costly.

Where to Measure: Context

Questions also arise about where one can test (context). As mentioned earlier, the setting for the research becomes a challenge when the target audience has dispersed immediately after the treatment. In the case of museums and community-based projects, formative research is relatively straightforward because one can find museum visitors and enrolled participants on-site. However, the target audience may never return to the site once the visit or project is completed. With television, the challenge is even greater. It is difficult to recruit the widely dispersed at-home audience in the first place to even know if there was any "treatment" and whether the treatment was the same for all subjects. Did some watch part of a program or all of it? Was there any consistency in the combination of programs viewed across subjects? Did some have greater distractions in the home setting, thereby missing key parts of programs?

It is not uncommon to recruit audiences to view a television program in order to ensure exposure, but this does not reflect the voluntary nature of informal learning. Despite these limitations formative research is often highly predictive of subsequent program performance in the television as well as the museum fields, as discussed in Chapters Two and Three.

Another difficulty in measuring informal learning outcomes is the ability to attribute changes in learning to the treatment itself, especially if a study attempts to measure changes over time. Did Cindy's knowledge of whales link back to her classroom experience, aquarium visit, or some combination of all of them? Can we parse out where her learning occurred? School-based evaluation faces this challenge as well, often neglecting to point out that learning gains (or declines) that occur across the school year are as much a function of what happens at home as in school. In fact, schools are sometimes more willing to assign *lack* of learning to the home environment, while learning *gains* are credited to school experience.

How to Measure: The Research Tools

In Chapter Three, Bitgood Serrell, and Thompson call for the development of research procedures which are unique to informal learning and are, at the same time, sensitive to the important goals of motivating target audiences. There is no question that measuring motivation to learn, for example, is a difficult task.

The tools that are most commonly used include observation (particularly meaningful for on-site evaluation), interviews (particularly appropriate for informal learning), and paper and pencil questionnaires and tests (desirable with large groups of participants). Within the informal learning field, there are proponents of face-to-face interviews and those who favor self-administered questionnaires. Some use both approaches. These will be discussed at greater length in each of the next three chapters.

Evaluating the Informal Learning Environment

In considering a summative research design, it is important not to limit the investigation to measuring outcomes or determining whether project goals have been met. It is important to think about other environmental factors that affect the success or failure of any project. In this way, practitioners and researchers can learn more about how to make informal learning experiences work rather than continue to make the same mistakes over and over again.

An evaluation of these environmental factors would include looking at the appropriateness of the format and distribution for the project. Is a television program the most appropriate format for a science topic that requires interaction and a museum exhibit the best place to transport one to far off lands? Is a documentary program appropriate for reaching at-risk science audiences such as females, minorities, or those with less than a college education when the audience ratings data are clear that a science documentary often attracts 35- to 49-year-old males who are white and college-educated? Is a museum exhibit appropriate for minorities if the museum seeking funding attracts almost no minorities to its institution? Is a children's program for 8- to 12-year-olds best placed on Nickelodeon, which regularly attracts that age group, or on PBS, which performs better with preschoolers and adults?

While schools do not need to promote and market their services, informal learning projects do. Yet informal science learning projects rarely obtain funding to communicate the availability of these projects to the public. This factor, however, must be considered in evaluating the success of each informal learning project.

In addition to distribution, format, and promotion, projects also should be evaluated in terms of their cost and packaging. Again, informal learning experiences compete for the attention of the public in an increasingly cluttered marketplace. To what extent did a project take this into account? If a television program has a lengthy five-minute introduction to the topic it is going to cover, its ability to hold the audience at the outset is threatened. The reality of the viewing environment is that viewers will decide within 30 to 60 seconds if they want to stay tuned to a program.

While these environmental factors do not, on the surface, seem germane to learning outcomes, they are, in fact, often as much the cause for projects falling short of their mark as is their content design. And, therefore, they need to be part of what we examine and evaluate as we look at the impact of informal learning.

Impact Research in Informal Science Learning

The purpose of this book is to examine those studies which determine the impact of informal science and math learning on targeted audiences. Toward this end, three authors were asked to write chapters which first defined informal learning in one of three fields: television, museums, and community-based projects. The goals of this book are to examine:

1. what studies have been conducted that measure the affective, cognitive, and behavioral impact of informal math, science, and technology projects;
2. what is known about what makes informal math, science, and technology programs work as a result of this research;
3. what research models and techniques are effective in measuring informal learning and what measurement challenges exist for informal learning; and
4. what next steps should be taken in the evaluation of informal learning in the areas of math, science and technology.

In Chapter Two, Dr. Milton Chen characterizes the informal learning environment for television, shares studies of the impact of television on science and math learning, and provides insight into what challenges lie ahead as we consider television as an informal science learning experience.

The second informal science learning environment, the museum setting (including technology centers, aquariums, and zoos), is discussed in depth by Dr. Stephen Bitgood, Beverly Serrell, and Dr. Don Thompson in Chapter Three. The museum experience differs from television in that it requires a commitment to leave the home and visit a center for a period of time (one to three hours on average, Falk & Dierking, 1992) and also in the degree of interactivity possible between the medium and the learner. The science experience moves from two dimensions (television) to three dimensions (exhibits, demonstrations), which can be interactive. The distractions change as we move from the privacy of the home with its interruptions of family life into public places where crowds are part of the learning environment. Bitgood, Serrell, and Thompson discuss these challenges and review the literature on the impact museums have on the public.

The third type of informal science learning experience is participation in community-based projects offered by organizations such as Girls Incorporated and is described in Chapter Four by Dr. Heather Johnston Nicholson, Faedra Lazar Weiss, and Dr. Patricia Campbell. Community-based projects often move beyond the single-shot experience which characterizes many museum visits and enroll students in a course over a period of time. Community-based projects, like some of the television projects reviewed by Dr. Chen in Chapter Two, target girls and minorities as important audiences. Since many community-based projects can structure learning experiences over time, some are quite rigorous in their expectations and goals for their students.

The final chapter of the book, written by Dr. Valerie Crane, is based on interviews with more than two dozen professionals, including informal science practitioners (science museum directors, exhibit developers, television producers, community-based project developers) and researchers in each of these fields. (The interviewees are listed in the Foreword of this book.) These interviews were designed to reflect on the future direction research and evaluation should take in the field of informal learning.

Each of the authors conducted an extensive review of the literature in his or her respective field. In addition to the authors' literature searches, Research Communications Ltd. conducted a mailing to over 300 professionals in search of any other articles that might indicate impact of informal learning. The impact studies sent to Research Communications Ltd. and those annotated by the authors are included at the end of this book. The audiences for this book include policy or decision makers, informal learning practitioners, researchers, and funders.

Summary

This chapter began by making the case for informal learning and establishing why informal learning is an important part of science education and research. The goals of this book are to examine what impact informal science learning experiences have had on the public through television (see Chapter Two by Chen); museums, zoos, and aquariums (see Chapter Three by Bitgood, et al.); and community-based projects (see Chapter Four by Nicholson et al.). The final chapter of the book presents the views and opinions of over two dozen professionals in the field gathered from interviews by Crane. Formative and summative research and evaluation on informal science learning were defined and challenges faced in summative research and evaluation were discussed. These themes are discussed in greater detail in the remainder of this book.

References

Bloom, Benjamin (Ed.). (1985). *Developing talent in young people.* New York: Ballantine.

Falk, John & Lynn Dierking. (1992). *The museum experience.* Whalesback Books, Washington D.C.

Miller, Jonathan D. (1987). *Scientific literacy in the United States.* In Communicating Science to the Public. Ciba Foundation Conference. John Wiley & Sons: New York. pp. 19–40.

Research Communications Ltd. (1987). *An exploratory study of* 3-2-1 Contact. Dedham, MA.

Research Communications Ltd. (1989). *An examination of* Square One TV as an informal math learning experience. Dedham, MA.

Suter, Larry (Ed.). (1993). *Indicators of science and mathematics education.* Division of Research, Evaluation and Dissemination, Directorate for Education and Human Resources, Washington, D.C.: National Science Foundation.

---

*The terms research and evaluation are used throughout the book and warrant working definitions. Evaluation studies focus on the effectiveness or impacts of informal learning efforts while research is used in broader terms to cover those studies which look for the common elements across informal learning settings which shed light on the nature of informal learning itself. Some studies described in this book make contributions both in the areas of evaluation and research.

2

CHAPTER

Television and Informal Science Education: Assessing the Past, Present, and Future of Research

Milton Chen, Ph.D.

> *If you are looking for ways to catch the close attention of children . . . show them, for starters, a small bird or a monarch butterfly or a nest of mealworms. To teach science, and to want to learn science, you have to have in mind some questions beyond answering, questions that raise other questions. Television is made for this kind of display, it seems to me, carrying the mind down to the deepest and smallest parts of nature, and out to the farthest reaches of the cosmos. But it should start with things that wiggle.*
>
> —Lewis Thomas (1984)

> *Most people, most of the time, learn most of what they know outside of the classroom.*
>
> —George Tressel (1990)

The Domain of Television in Informal Science Education

As George Tressel has argued, the process of individual intellectual development and lifelong learning is largely based on experiences and encouragement that come from outside of formal schooling. And as Lewis Thomas has observed, television has the potential to play an important role in that process, from early infancy through adulthood. This chapter examines the research to date on the role of television in informal science education. The impetus for this chapter and this book comes from the National Science Foundation, which has made a substantial investment in the broadcast media's ability to reach a large, national audience with science programming, dating back to the first season of *NOVA* nearly 20 years ago.

During four fiscal years, 1987 through 1990, NSF spent more than $25 million on the production and broadcast of 20 television series on science and technology topics (National Science Foundation, 1991). Three children's series, *Square One TV* and *3-2-1 Contact*, produced by the Children's Television Workshop, and *Reading Rainbow*, produced by Nebraska Educational Television, Lancit Media, and WNED, received the majority of funding, more than $20 million. The remaining $5 million funded a diverse set of programs, from WQED's *The Space Age* to WNET's *Childhood* and *Medicine at the Crossroads* and WGBH's series on molecular biology and the history of the computer. An additional $1 million was spent on science coverage on National Public Radio.

Appendix 1 contains a selective listing of about 50 of the major TV series broadcast over the past 20 years on science, mathematics, and technology. Historically, the majority of these programs have been broadcast on PBS. For adults, these series include *NOVA, The Ascent of Man, Cosmos, National Geographic Specials, Nature, Connections* with James Burke, *Discover: The World of Science, Newton's Apple,* and more recently, *Scientific American Frontiers* and *The New Explorers* with Bill Kurtis. For children, productions have included *3-2-1 Contact, Square One TV, Voyage of the Mimi,* and the science-related programs in *Reading Rainbow.* The Annenberg/CPB Project has also funded a number of series intended for both prime-time viewing as well as adult telecourse use, including *Race to Save the Planet* on the environment and *For All Practical Purposes* on mathematics.

Commercial networks also have broadcast science programs, which have included *The Undersea World of Jacques Cousteau, Mr. Wizard, Walter Cronkite's Universe,* and *Wild Kingdom.* While the amount of program-length science programming on commercial broadcast networks has declined through the 1980s, commercial stations continue to carry science news reports on both local and national nightly news programs and morning talk shows.

The past decade has seen a tremendous growth in the distribution of video through channels beyond broadcast television. This expansion of technological options has affected the availability of science programming as well. Beginning in the 1980s, cable television also established a reputation for science programming, primarily via *The Discovery Channel.* Series that had previously been broadcast on PBS have been carried on *The Discovery Channel,* and the Channel has produced original science productions of its own. Other cable services also carry science-related programming, such as CNN's *Science and Technology Week* series. In the early 1980s, with the diffusion of VCRs, home video has also become a distribution mechanism for programs related to science, with programs such as the *National Geographic Specials* available in home video stores for rental and purchase.

Other technologies for carrying video material on science are becoming more widely available, but have yet to show mass penetration in homes and other informal environments. These technologies include videodisc (the *National Geographic Whales* and *NOVA: Animal Pathfinders* are among the best known titles), compact disc, and satellite delivery. As video and computer technologies merge through this decade, existing science video material will be "repurposed" and reformatted for these new media, such as a new Nature Interactive multimedia system currently in development, based on the *Nature* series with George Page. New TV series on science, mathematics, and technology continue to be produced, often anticipating distribution beyond broadcast, through videocassette, videodisc, CD-ROM, and online computer networks. Now is an opportune time to take the measure of work done in this field during the previous two decades, to inform planning and funding efforts for this new media environment, and to assess the foundation upon which future research should build.

A primary aim of this chapter is to summarize the research in the field of television and informal science education. A striking initial observation is the mismatch between the level of national funding and production effort and the paucity of systematic research on these projects. From a research perspective, the field has gone largely unexamined. As a review of research studies will show, the main body of research relates to children's science series funded by NSF and produced by the Children's Television Workshop in New York. Very few of the science series produced for a general adult audience have been the subject of research or evaluation.

The reasons for the lack of a substantial research base are many, but can be traced to several fundamental conditions:

1. *Methodological challenges confronted by research "in the field" on television and informal science learning.* Those who conduct research on the educational effects of television programs on science confront some difficult methodological challenges that do not yield easily to the standard approaches of social science research. In

studying outcomes, this genre of research does not fit easily into the experimental paradigm. For example, the "treatment" is highly variable. The television "stimulus" is complex and evades simple categories and schemas. The "sample" for any single program is highly diverse, consisting of national audiences of viewers who vary in their age, geographic location, experiences with the topic of the program, educational and work background, and other informal learning experiences. The community of researchers who work on these problems is small, and would benefit from greater opportunities to share their work with each other and the larger academic community.

2. *The complexity of the "home viewing environment."* This environment is actually many environments, which differ by types of households and families. The television program must vie for viewers' attention with multiple distractions, competing activities, and, in the age of the remote control, numerous channels only a click of a button away. The "rational" image of a single viewer selecting a science program to watch, sitting down on a couch in front of the set at the start of the show, and watching a complete program from start to finish, is even less applicable in the 50-channel home than it was in the pre-cable era. It is also unlikely that viewers understand the information presented in the program in precisely the way intended by the program's producers. For these reasons and others, research on learning from science programming also has not addressed issues of learning over the longer term. Some research has studied short-term cognitive and affective change, but longitudinal effects, not just of several weeks or months, but over several years, have yet to be investigated.

3. *The "lack of transfer" between much of social science research in psychology, communication, and education to the specific research questions faced by the field.* Much of the groundbreaking work in cognitive and social psychology during the 1980s offered theoretical insight into, for instance, the role of domain-specific knowledge in the development of expertise, or the importance of learning in small groups through peer tutoring. While this work has important implications for learning from television and other media, these theoretical bridges have yet to be built. It is likely, for instance, that television material can contribute to viewers' knowledge in specific scientific or technological domains, and/or that informal discussion of television programs can support a viewer's continued learning and motivation.

4. *The "producer-driven" nature of television projects in science educa-tion.* Generally, the project directors for these projects have tended to be producers, rather than science educators or television research-ers. While producers vary in their interest in and understanding of evaluation research, it is fair to say that most television producers and producing organizations have not placed such research on their programs as a priority. Notable exceptions include the Children's Television Workshop and other producers, such as WGBH-Boston, for whom formative evaluation has been incorpo-rated as part of the producing culture and encouraged by funding sources, such as the National Science Foundation.

5. *The historical absence of evaluation research, both formative and summative, in funding requirements and project milestones.* In order for research in this field to attain higher status and greater vitality, it must be incorporated into project funding and develop-ment at the earliest stages. Since the 1970s, many series have been funded and produced with little or no evaluation of any kind. The complementary value of formative research, to aid the develop-ment of the most effective program, and summative research, to assess its impact, is not widely understood in the science television community.

The next section, describes the size and nature of audiences for science programming for broadcast and cable TV. Information on audi-ences for science programming on radio is also given, acknowledging that, similar to television, the role of radio in promoting public under-standing of science has not been systematically studied. Audience data for radio are given mainly to illustrate their growing size and the range of science programming available. The section entitled Review of Research Studies summarizes findings from those studies that have investigated "effects" or "outcomes" from viewing of science TV programming. The final section of this chapter presents commentary on the serious need for a stronger program of research on the role of video and audio media in informal science education and recommends future directions for this research.

Audiences for Science Programming

Broadcast Television

One of the main arguments for utilizing television as a tool for promoting public understanding of science is its reach. Television is available in

98% of households, a figure that has held steady for the past decade. With 92 million TV households in the country, broadcasts of science programming on PBS can reach several million households over several weeks. Science series such as *NOVA, National Geographic Specials, and Nature* typically are among the most-watched programs in all categories of public television programming. The reach of science programming is illustrated in Table 1 below, summarizing PBS audience data for science series broadcast during the 1991–92 year (Public Broadcasting Service, 1993).

TABLE 1: National Audiences for Selected PBS Science-Related Programming, 1991–1992, for Households and Persons

| | | | Households | | | | | Persons | |
| | | | Cumulative Audience | | Average Audience | | Avg. Mins. | Cumulative Audience | |
	# Eps.	% Cov.	Rtg. %	(000)	Rtg. %	(000)	Viewed	Rtg. %	(000)
Machine That Changed The World	2	95	4.0	3,690	2.5	2,260	35	2.0	4,640
National Audubon Specials	4	89	3.5	3,230	2.0	1,870	33	1.9	4,520
National Geographic Specials	4	96	8.3	7,670	5.6	5,110	39	4.7	11,050
Nature	7	88	5.1	4,660	3.0	2,800	35	2.8	6,620
NOVA	10	92	6.0	5,510	3.7	3,360	38	3.1	7,270
Scientific American Frontiers	1	88	4.7	4,330	2.5	2,300	30	2.4	5,700
Square One TV	12	86	2.1	1,920	0.5	440	28	1.1	2,610
3-2-1 Contact (Series)	10	57	1.1	1,010	0.2	210	24	0.5	1,260
3-2-1 Contact Extra (Special)	1	94	2.9	2,670	1.7	1,570	33	1.4	3,330

NOTE:

"# Eps." represents number of episodes;

"% Cov." represents percentage of U. S. TV households covered by the broadcasts;

"Cumulative Audience" indicates households or persons reached by broadcast of the multiple episodes of a series, not including repeat broadcasts;

"Average Audience" indicates percent of households tuned to an individual show in a series, on average;

"Rtg. %" indicates series rating as a percentage of U. S. TV households;

"Avg. Mins. Vwd." stands for average minutes viewed per episode. Episodes of adult series are 60 mins.; *Square One TV* and *3-2-1 Contact* are 30 mins.; the *3-2-1 Contact Extra* was 60 mins. long.

The *National Geographic Specials,* traditionally among the most highly rated TV programs on PBS, continued to attract large national audiences, with four *Specials* that year being viewed in 7.67 million households (8.3% of households) by more than 11 million individuals (4.7% of Americans). Programs broadcast that year included "Hawaii: Strangers in Paradise," "Eternal Enemies: Lions and Hyenas," "The Mexicans: Through Their Eyes," and "Braving Alaska." These data address only the premiere broadcasts of the *Specials* that year; the cumulative number of households and persons reached would increase with subsequent repeat broadcasts. Average minutes viewed for the hour-long programs was 39 minutes, indicating that, on average, viewers' sets stayed tuned to two-thirds of each program.

Audience figures for other series, such as *NOVA, Nature,* or *Scientific American Frontiers,* while lower, also demonstrate the millions of households and individuals reached by science programming on broadcast television. More specific data on *Square One TV* and *3-2-1 Contact* are discussed below in relation to their target audience of 8- to 12-year-olds (6- to 11-year-olds is the closest Nielsen age category). While total national household and persons data are less meaningful for series targeted for a narrow age category, the data on a *3-2-1 Contact* special on sex education, intended for both parents and older children, nonetheless demonstrate the tremendous reach of a single program, viewed in 2.67 million homes.

Chart 1 below illustrates variation in audience ratings for one series, *NOVA,* during the 1991–92 season. The chart gives ratings for individual *NOVA* episodes that year, for the 25 major markets providing ratings overnight through Nielsen audience meters. The universe of TV households for that year was 92.1 million. One ratings point, or 1%, equals 921,000 households. The average *NOVA* rating for the year, 3.1, translates into an average of 2.85 million households viewing a *NOVA* episode. Of interest are those programs with the highest ratings ("Who Shot Kennedy", "Hell Fighters from Kuwait," and "Submarine!") versus those with the lowest ratings ("What Smells?" and "Skyscraper!"). These titles reveal a strategy on the part of *NOVA* producers to attach more provocative titles to programs, as one technique for piquing viewers' interest and bringing them to the broadcasts. The data also suggest that some topics may be inherently more intriguing to the public, e.g., a show on forensic evidence related to the Kennedy assassination versus one on how humans perceive smells.

When children's series are broadcast on a daily basis (5 programs per

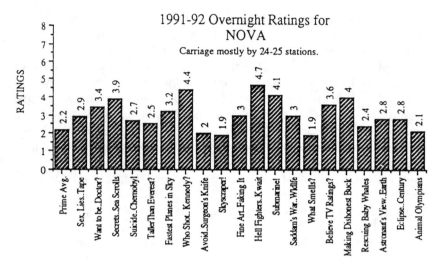

CHART 1: National ratings for *NOVA* episodes, 1991–92, in 25 major markets

week with some weekend repeats), the multiple broadcasts can generate a larger cumulative audience among the specific target-age category of viewers (e.g., 6–11s) than is typical for a weekly series. Representative data on *3-2-1 Contact* come from a study of national Nielsen ratings for its second season, broadcast from January to March, 1984. With the first season in repeats since its premiere broadcast in 1980, public awareness had several years to develop.

3-2-1 Contact typically averaged a national Nielsen rating of about 1, or about 1% of all TV households viewing during a typical program. Using the universe of 83 million TV households in 1984, about 830,000 households watched an average minute of a program. Controlling for households with children, the percentage figure rose to 2.2% of households with 6–11s and 4.5% of households with preschool children ages 2 to 5.

This latter figure demonstrates one of the challenges of reaching child audiences via broadcast: the strong effect of audience flow and lead-in programming. The audience for children's programs on PBS included a majority of preschool children who were at home and drawn to watching *Sesame Street,* which often preceded *3-2-1 Contact.* As a result, many preschool viewers "flowed" into the viewership for *3-2-1 Contact.* PBS preschool programming, aided by *Barney & Friends,* continues to draw large numbers of preschool viewers, presenting challenges for bringing older children to programs such as *3-2-1 Contact* and *Square One TV.*

Since *3-2-1 Contact* was typically broadcast Mondays through Fridays over a period of weeks, cumulative audience figures show an increase in the numbers of homes with children watching over one week and eight weeks, as compared with a single episode. In one week, 10.1% of households with a 6- to 11-year-old watched at least one program. Across an eight-week period, 33.2%, or about a third of households with a 6- to 11-year-old, watched at least one *3-2-1 Contact*, translating into 4.2 million households. The figure was slightly higher among households with 2–5-year-old children (36.6%, or 4.7 million households).

Some data were available on cumulative audiences in 1984 by households with children and income, or education. Comparing only households with children, a somewhat higher percentage of households with incomes over $20,000 (31.5%) or some college education (28.7%) watched *3-2-1 Contact*, compared with households with incomes less than $20,000 (19.2%) or educational levels at a high school degree or less (24.0%).

Ratings for *Square One TV* showed many of the same patterns as those for *3-2-1 Contact*. Ratings were again higher among households with preschoolers than households with 6–11s. Audiences were highest for the first and second seasons, 1987 to 1989, declining in subsequent years due to less promotion and carriage on fewer PBS stations. In its premiere year, 1987, *Square One TV* averaged a weekly cumulative audience of more than 3.9 million households. As Table 1 indicates, five years later, the weekly audience had declined to half, or 1.9 million households.

Cable Television

During the 1980s, cable penetration experienced a rapid growth in American homes. By late 1992, according to the A. C. Nielsen, 61.5% of American households were cable subscribers (Public Broadcasting Service, February, 1993). Cable subscription is highly correlated with household income. Among households with less than $10,000 income, 47% subscribe to cable. For households with greater than $60,000 income, 80% are cabled. The rise of cable television has greatly expanded the number of TV channels available in homes. Pooling cable households with non-cable households, the average number of channels received by American households rose from 9 in 1981 to 38 in 1992.

The Discovery Channel has established its identity as the cable service offering programming on science, nature, and technology. The Boston Consulting Group (1991) reported that, in 1990, among cable viewers, 41% are more likely to turn first to *The Discovery Channel* for nature and science programming, compared with 33% who would turn first to public TV. Only three years earlier, 47% would have first turned to public TV, followed by 28% turning to *The Discovery Channel*. In total funding for programming, *The Discovery Channel* also is competing with public TV, spending $35 million annually for science programming, compared with public TV's $49 million. Audience data for *The Discovery Channel* were not publicly available.

As with cable TV, videocassette recorders (VCRs) achieved widespread popularity through the 1980s and provided another technology for expanding viewing choices.[1]

Radio

Science reporting and discussion programs also are found on radio. Some of the most regular reporting is done through National Public Radio (NPR), whose science programming has been funded by NSF over the years, e.g., Ira Flatow's *Science Friday*, which began airing in 1991 on NPR's *Talk of the Nation*. Recent *Talk of the Nation* programs, which air from 2 P.M. to 4 P.M. EST, have featured discussions with physicist Philip Morrison and his wife and science educator, Phylis Morrison, and inventor Jacob Rabinow. Within a few months of its premiere, *Science Friday* was carried in 40 markets, and by mid-1993, its broadcasts were carried on more than 60 public radio stations, including major markets such as New York, Philadelphia, San Francisco, and Washington, D.C., as well as numerous small markets. Other science programs on public radio include *Star Date* and more recently, *Earth & Sky*, produced by Joel Block and Deborah Byrd at the MacDonald Observatory in Austin, Texas. The American Association for the Advancement of Science (AAAS) has produced brief "Science Update" segments broadcast regularly on the commercial Mutual Broadcasting System.

Estimates for audience size for specific programs are not readily available. It is clear, however, that audiences for public radio have been growing. From 1986 to 1991, audiences for National Public Radio, the

---

[1]Like cable, VCRs also experienced impressive adoption during the 1980s, along with falling prices for the technology. While 78% of all American households have at least one VCR, the technology has been more popular than cable with lower-middle and middle-income households. Among households with $20,000 or more income, 87% have a VCR. Among households with $30,000 or more income, the figure rises to 91%. At the lowest income levels, however, only 44% of households with $10,000 or less income have a VCR. While data on use of VCRs for viewing of science-related programming were not available, we can surmise that such usage is low, since most VCR usage is for rentals of home movies. These data are included here mainly to suggest the availability of this distribution mechanism for video material.

major provider of public radio programming, have increased 35%, from 10.2 million to 13.8 million listeners. These listeners tend to be better educated (two-thirds have attended college, compared with 20% in the general population), participate in elections (54% compared with 42% of the general population), and be more active consumers (27% had contacted media or written a company about a product, compared with 17% of the general population). As with data on TV viewers of science programming, it will be important to describe more specifically the population of radio listeners for science programs.

No evaluations of science programming on radio were located, either via literature search or phone conversations with the producers. A more comprehensive study of the public's exposure to science media should address the role of all media—television, radio, newspapers, magazines, and books—and should investigate the information-seeking patterns of individuals across these media. One finding, based on Jon Miller's research, is that about 20% of the population actively seeks science information using these media.

Review of Research Studies

This section reviews the relevant literature of studies on viewers' learning from television in informal science education. The methodology for locating these studies involved searching the ERIC system, supplemented with other references of both published and unpublished literature (e.g., research reports, conference presentations) not indexed in ERIC. Through a call for papers, the editor of this book supplied a number of unpublished reports. A number of books on topics related to informal science education and television were also reviewed.

The paucity of research studies on the effects of television in science education, even more specifically informal science education, was readily apparent in assembling the literature for this review. For instance, searching the combined set of studies with descriptors of "science education" and "television" in the ERIC system resulted in 90 references, only 15 of which are research studies reporting empirical data, some addressing instructional use, others reporting on formative research data. Narrowing the field specifically to those studies reporting "effects" types of data, together with other studies recommended to the author, produced the list in Table 2.

The shortage of research is apparent both in the total number of studies, as well as the absence of much recent research on the topic. When compared with the growth of research on science learning in

TABLE 2: Studies of Television & Informal Science Learning, by Children, Adolescents, & Adults and Mathematics or Science Field

	Mathematics	Science
Children, Ages 6–12	Hall, Esty & Fisch (1990) Research Communications (1989)	Chen (1983) Johnston & Ettema (1982) Johnston & Luker (1983) NFO Research (1990) Ormerod, Rutherford, & Wood (1989) Research Communications (1987a) Rudy (1981)
Adolescents, Ages 12–17		Research Communications (1992)
Adults, 18 and Over		Gagne & Burkman (1982) National Educational TV (1969) Research Communications (1987b) Public Broadcasting Service (1987)

museum settings, the low volume of research activity should be a matter of concern for researchers, practitioners, and policymakers. It will be difficult to lay a foundation for the continued development of theory and methodology in the absence of a vital research effort.

For instance, not one study could be located on viewers' learning from *NOVA* or the *National Geographic Specials,* among the best known science and nature programs on television for decades. Nor has there been any systematic study of viewers' learning from the wealth of other science programming carried in recent years on the PBS system, as well as cable channels such as *The Discovery Channel.* Likewise, we know little about how viewers learn from frequent science-related news reporting on television and radio. Since audience ratings suggest that viewers who watch one science program tend to watch others and comprise a "niche audience" for science programming, more comprehensive study of the "science viewing habits" of this audience would be one obvious area for future research.

The chart on the following page summarizes the literature located, by author, year, and three categories of audiences—children 6–11, adolescents 12–17, and adults 18 and over. Some of these studies collected data from students in school settings, but did not include any teacher training or utilization of the video material in class. The classroom served only as a convenient location for groups of viewers. Since these studies involved "viewing only" within a classroom, they are included here as studies addressing informal learning.

Children, 6 to 12

Among all TV series, the Children's Television Workshop productions in science and mathematics education, *3-2-1 Contact* and *Square One TV*, both for 8- to 12-year-olds, have generated the most research activity on potential cognitive and affective outcomes. *3-2-1 Contact*, a magazine-formatted show, combined short documentary segments hosted by young role models with a dramatic mystery serial, The Bloodhound Gang. *Square One TV* emphasizes making mathematics entertaining through comedy sketches, songs, and a repertory company of adult actors. Mathnet, a takeoff on *Dragnet*, featured two police detectives solving problems using mathematical thinking. In formative research, both The Bloodhound Gang and Mathnet were the most appealing parts of their programs, owing to their dramatic plots and character development.

3-2-1 Contact (broadcast life: 1980–1991)

During the second season of *3-2-1 Contact*, two studies were commissioned by CTW to investigate potential cognitive and affective effects of the series (Chen, 1983; Johnston & Luker, 1983). The studies were similar in several respects: children viewed 3-2-1 Contact every day for several weeks in classrooms without teacher intervention. The study by Chen involved four mixed classrooms of 4th- through 6th-graders (N = 101) at an elementary school in Oakland, California, which contained 70% minority children. The study by Johnston and Luker used a sample of 192 4th- and 5th-graders in a predominantly white, suburban Michigan school. Children in both schools were from lower to middle SES groups.

Both studies found that children acquired a surprising amount of scientific information from the programs, ranging from the science behind athletic equipment, such as the design of tennis shoes and racing suits, to the design of airplane wings and the development of a fetus. For most viewers, *3-2-1 Contact* represented their first encounters with these phenomena and concepts. Consequently, children acquired an awareness of a wide range of phenomena, the beginnings of key information and concepts (e.g., the name of a bird, "hummingbird"; its appearance; its behavior; that it can beat its wings at the rate of 60 times per second), and, perhaps most importantly, an interest and curiosity in learning more.

As Johnston and Luker found:

> . . . children clearly learned and retained a lot of science information from their two weeks of viewing. This includes labelling (DNA) and more complicated discriminations such as knowing what can and

cannot be cloned, and why speed skaters wear racing suits. Most impressive, however, were the changes in children's ability to explain satisfactorily an abstract—and important—principle such as Bernoulli's explanation for how winged objects fly. The fact that this was accomplished with a show that is more of the genre of entertainment than instructional television is especially notable.

The finding of the amount and range of scientific knowledge absorbed by children poses an interesting challenge for the science program designers and producers, who, as with *3-2-1 Contact*, often target changes in motivation and affect as their primary goals. Chen hypothesized that increasing the knowledge base of young children, both in depth and breadth of phenomena, may be an important prerequisite to an overall shift in motivation and attitude towards science. It may be that children who acquire negative attitudes about science may be basing those beliefs upon only superficial experience. Conversely, children who indicate an interest in science must have some elementary knowledge and experiences upon which to base their interest.

Children and teachers in both studies also indicated that *3-2-1 Contact* motivated an interest in learning more about science (e.g., reading a book, doing an experiment, making a model). The concept of the "teachable moment" is relevant here. *3-2-1 Contact* presented a wide range of phenomena and helped to trigger children's curiosity about them. Children who viewed these programs were ready to learn more, but whether families, schools, and other informal experiences were provided to continue their learning was unclear from these studies.

This may be one of the chief effects of television—to whet the appetite for science, to serve as a springboard for other activities conducted in the home, the classroom, the after-school setting, or the science museum. Ensuring that these activities are widely available to children remains a major challenge for the field of science education.

The series also appeared to help children shift to a broader concept of science and its topics of study.

Some viewers were able to note shifts in their own feelings about science before and after viewing the programs. These comments generally noted a change away from a perception of science as 'boring' and requiring much study and work in labs to a sense that science can be fun. Accompanying this shift was a broadening of topics that fit under the label of "science" (such as scientific aspects of sports). (Chen, 1983)

Four years later, Research Communications Ltd. (1987) conducted a national survey of children and parents in 31 markets around the country to measure awareness, viewing, and impact of *3-2-1 Contact*. Among the 410 parent-child pairs interviewed in their homes, with children ranging in age from 4 to 12, awareness and exposure (defined as any viewing) were high. Among parents, 70% had heard of the series; 73% of children reported awareness. About 60% of children and parents reported some viewing of the program. Frequency of viewing was estimated at less than two to three times per month.

This study reported that children younger than the intended target audience of 8- to 12-year-olds were watching *3-2-1 Contact*. More than half (54%) the parents reported that their children began watching the series at the age of four or younger; only 17% of parents said their children began viewing at the age of eight or older. As suggested earlier, since many PBS stations carried *3-2-1 Contact* as part of its block of children's programs, which has historically been dominated by *Sesame Street* and *Mister Rogers' Neighborhood* preschool series, many preschool-aged children were exposed to *3-2-1 Contact*. This finding from the Research Communications Ltd. study provides independent confirmation of Nielsen ratings data.

Self-reports from children indicated that they viewed the series because it was "interesting" and they "learned new things." Barriers to more frequent viewing included other competing activities and TV series. The researchers noted that, while children did report learning from the series, the younger audience skew resulted in many younger-than-intended viewers (4- to 7-year-olds) indicating a higher level of difficulty with program content.

In a key finding from this study, 61% of children carried out a science-related activity as a result of viewing. Consistent with the study by Chen (1983), a majority of children followed their viewing by discussing it with a friend, teacher, or family member, conducting an experiment, visiting a museum, or reading a book. The hypothesis that children's TV series on science can trigger interest in other learning experiences gains further support from a study of children's book reading following *Reading Rainbow* (see below, NFO Research, 1990).

Children's awareness and viewing of 3-2-1 Contact was correlated with parents' education and income levels, viewing of public TV, and science interest. The researchers note, however, that "there are still significant numbers of children viewing who have parents with lower education, income, PTV viewing, and science interest."

A related study from Great Britain also found correlations between science interest of 10- to 13-year-olds and viewing of the popular science series (Ormerod, Rutherford, & Wood, 1989). Frequency of viewing of science programming was correlated with interest in science, among both girls and boys. The study also investigated gender differences in science topics covered by the programs, and found that middle-school boys preferred programs on space, while girls preferred programs on animals, a finding consistent with earlier formative research for *3-2-1 Contact* (Mielke & Chen, 1983). That research recommended that further attention be given to message design elements, such as gender of presenting cast members and emphasis on human versus technological dimensions of the topic, in order to increase the appeal of topics across gender.

These topics were further studied in a doctoral dissertation on gender-specific reactions to *3-2-1 Contact's* science topics and young cast members. Rudy (1981) utilized a field-experimental design to study boys' and girls' attitudes towards the young role models in *3-2-1 Contact,* using viewing of 13 programs as the treatment. While both boys and girls increased in their level of science interest after viewing, the initial level of interest was a key predictor of positive reactions to these role models, for both boys and girls. Boys preferred and remembered more about the male cast members, while girls had greater interest and recall for the female cast members. Girls who exhibited less stereotypically masculine traits, as measured by two sex-role attribute scales, responded more positively to the female cast members.

It is interesting to note, however, that the female cast members declined in their appeal ratings over the course of viewing, suggesting that the personal qualities of role models are critical to maintaining interest. Since this study was conducted on programs from the first season of *3-2-1 Contact,* and substantial attention was devoted to improving the casting and attributes of subsequent cast members, this area of research, on the design and effects of programming by gender, continues to be ripe for further research.

Square One TV (broadcast life: 1987–present)

Square One TV has been the focus of two studies that begin to demonstrate some convergence of findings regarding its value in an informal viewing setting. Both studies examined the effects of *Square One TV* for children viewing alone, without adult mediation. In one study, a test group of children was asked to view the series as often as possible, under natural conditions in their own homes, during a four-week period (Research Communications, 1989). Pre- and post-tests of problem-solving ability were administered, with comparisons made to a control group of non viewing children. These measures consisted of one-on-one interviews subsequently coded for content.

No difference was found in problem-solving ability between experimental and control groups, interpreted largely due to low frequency of exposure to the program. Average viewing was five programs during the eight week period, less than one program per week. (It was unclear whether viewers watched complete programs or partial programs. Since the viewing was voluntary, viewers were credited with viewing of a program based on partial viewing.)

This study focused especially on "cognitive mediators" that might promote effective problem-solving, defined as task enjoyment, task persistence, and effectiveness of strategies. The researchers noted that, even though such mediators were not specifically emphasized in an overt fashion by the show's characters, some gains on these factors were evident. Compared to nonviewers, viewers tended to demonstrate greater task persistence and more effective strategies.

In contrast, the second study, conducted by the CTW research staff (Hall, Esty, & Fisch, 1990), mandated a high amount of *Square One TV* viewing. (Viewing for this study was done in a school, but without any teacher mediation.) This study also occurred over eight weeks with a viewing sample of 24 children watching 30 episodes, *six times the amount of viewing in the first study, close to one program per day (complete programs were viewed).* This study also developed innovative measures of problem-solving, where hands-on, nonroutine mathematical problem-solving tasks were presented to experimental and control groups of children at pre- and posttest. Children's performances were videotaped and coded by observers who were not present at the schools. This study is unique in developing a new methodology for measuring a richer set of outcomes than can be measured by paper-and-pencil instruments alone.

> Viewers showed statistically significant gains in the number and variety of problem-solving actions and heuristics they used to solve problems (e.g., working backwards, looking for patterns). Viewers showed significant gains in the mathematical completeness and sophistication of the solutions that they reached.

Viewers also demonstrated more positive affective outcomes regarding their motivation and interest in mathematics, willingness to pursue more difficult and challenging problems, and enjoyment of mathematics.

These findings are impressive and suggest that a high frequency of viewing, without further mediation and support from adults, can result in positive cognitive and affective growth in mathematical thinking and interest. Further research should build upon this study to replicate its findings with a larger sample and to analyze subgroups by sex, age, socioeconomic status, and ethnicity.

Reading Rainbow (broadcast life: 1983–present; science programs, 1986–present)

Reading Rainbow, a half-hour program hosted by LeVar Burton, presents children's books, book reviews, and related documentary segments. Originally developed as a summer series to help combat "summer reading loss," *Reading Rainbow* has become a favorite among elementary-school teachers. In the most recent national survey of school use of instructional video, *Reading Rainbow* was the most popular series in the nation's classrooms, used by a projected 132,600 teachers with 4.2 million students (Corporation for Public Broadcasting, 1992).

In order to prepare for year-round *Reading Rainbow* broadcasts, including those science-related programs funded by NSF, a national sample survey was conducted (NFO Research, 1990). A total of 707 parents of 5-through 8-year-olds responded to a four-page questionnaire of items concerning familiarity with and attitudes towards *Reading Rainbow*, family reading habits, and parental involvement with children's reading, such as encouragement of reading for pleasure.

The findings support the view expressed above of the importance of parental support for viewing of an educational series and the participation in related activities, such as reading of storybooks. Familiarity with the series was high (70%) among parents, although lower among single parents (52%). Familiarity was highest among parents who read themselves, as measured by reading of newspapers, books, and magazines. Of those aware of the series, 72% encouraged their children to watch.

One "effect" of the series appears to be to motivate children to read the books seen in the programs, also a finding consistent with the research from *3-2-1 Contact*. Over half of parents familiar with *Reading Rainbow* reported that their children asked for the books featured, with boys and girls equally likely to do so.

One of the most revealing findings is that parental encouragement to view the program was most highly correlated with the child's requests for the books, suggesting again that a strong family environment was at work in encouraging, modeling, and approving important and worthwhile activities. Conversely, children who were not encouraged to watch the program were less likely to ask for books. Children also were more likely to ask for the books if they visited a library at least monthly; liked to read both at school and for pleasure; and lived in a family that reads regularly. For the purposes here, a study more closely focused on the science programs in *Reading Rainbow* would be more revealing regarding relationships with other informal science activities, such as children's asking of science-related questions, trips to science museums, or engagement in hands-on activities in the home.

Reading Rainbow also has conducted a survey of children's librarians in public libraries, supporting the finding that a main effect of an appealing children's science program is to motivate the reading of books. Three surveys have been conducted, in 1983, 1985, and most recently, 1988. In the most recent study, surveys were sent to a random sample of 800 librarians, and stratified by four geographic regions. While the librarians responded with strong positive reactions as to the value of the series and its presentation of positive role models, the most interesting findings addressed behaviors undertaken by themselves and the parents and children they observed.

Among this sample, 58% of the librarians provided information about *Reading Rainbow* books and the TV series to parents and 44% read the related books at story times. Over 82% reported that children asked for books seen on the program, with 42% recalling children asking specifically for featured science books. More than half of the librarians (52%) reported parents asking for books mentioned on the series. Book sales have done exceedingly well, with publishers reporting sales increases of 1.5 to 9 times the typical children's book, once a book has been featured on the series. Publishers of books for adults related to other nationally televised science series, from Carl Sagan's *Cosmos* to James Burke's *Connections*, report similar effects following broadcast. These findings support a hypothesis for further study, that TV viewing can promote other positive science learning behaviors, especially where such behaviors are clear-cut (books to be found in libraries are featured in the shows) and conveniently performed (proximity of library branches to families).

Adolescents, 12 to 17

FUTURES (broadcast life: 1990 to current)

FUTURES, produced by Foundation for Advancements in Science and Education (FASE), is a series of 15-minute programs designed to encourage interest in careers using science and technology. The series is hosted by well-known mathematics teacher Jaime Escalante, and includes interviews with scientists and engineers as well as a broad range of other professionals who utilize science and mathematics, such as Air Force pilots, fashion designers, and athletes.

Research Communications Ltd. (1992) conducted an evaluation of the series, using a field experimental design. Junior high students in seven schools located in four cities viewed 12 programs from the series during one semester, with a comparison group of nonviewing students. Measures of appeal and attitudes were taken before viewing, at the end of viewing, and one month later.

Findings compared 88 viewing students with 88 non viewers. *Futures* had a positive impact on student attitudes towards mathematics and science, although the effect was one of mitigating decline in interest with this target audience of 12- to 14-year-olds. Students who saw *Futures* had higher levels of agreement with statements such as "Careers that use math can be creative" or "Knowing mathematics will allow me to do some incredible and exciting things in my future career" after viewing the series and after a one-month period.

However, both viewers and non viewers experienced an overall drop in interest in mathematics and science over the semester. There was some evidence from this study that the series was especially appealing to African- and Hispanic-American students. The series does make a concerted effort to include role models from these groups, although it is unclear what relationship such role models have to the perceptions and attitudes of young people. An implication of this study is the need to devote further resources to the middle-school age group to reverse this decline in interest, especially as attitudes during this age form opinions about course-taking during high school.

Adults, 18 and Older

As noted earlier, there has been a disappointing lack of summative evaluation on the large volume of science programming broadcast on PBS, *The Discovery Channel* and other cable services, and commercial stations. The need to develop a body of research and evaluation on adult learning from science programming should be a priority for future funding in the field of informal science education.

This body of research will need to be built, theoretically and methodologically. Fundamental work, such as explicating major dependent and independent variables, remains to be done. For example, how might viewers categorize short-term effects of viewing science programming, along dimensions of learning or impact? For nearly two decades, researchers have attempted to describe the types of outcomes that might result from viewing of educational television. Public television researchers, in particular, have attempted to develop alternative ratings systems to the strict quantitative ratings delivered by the Nielsen or Arbitron systems (Corporation for Public Broadcasting, 1979). These alternative ratings systems, described as "qualitative ratings," utilize a variety of methodologies to gather data on viewers' reactions to the appeal, comprehensibility, dramatic impact, or educational value of programs. These methodologies have ranged from computer-based technologies by which viewers input responses through keypads to adaptations of the Nielsen diary method.

This latter method was used in a national random sample survey conducted by the Public Broadcasting Service (1987). For two weeks in the spring of 1987, more than 2,000 adults kept qualitative diaries of viewing of both commercial and public stations, logging stations and programs watched and responses on 10-point scales assessing program appeal, and the extent to which "I learned something from this program" (learning scale) or "This program touched my feelings" (emotional impact scale).

Programs on science and nature scored high on appeal and learning scales. Among 12 program types, science documentaries had the highest average appeal ratings (84). Four of the top 10 most appealing shows, among more than 100 listed, were science documentaries on PBS: two *NOVA* programs ("Animal Pathfinders" and "Are You Swimming in a Sewer?") and two *Nature* programs ("Lord of Hokkaido" and "Selva Verde"). Game shows, fantasy programs, and comedy shows had the lowest appeal ratings. Interestingly, the most popular programs on television were given the lowest appeal scores. While the numbers of viewers selecting science programs were fewer among this sample (about 50 to 75 viewers), this smaller sample of science-attracted viewers apparently felt more strongly about the appeal of these programs than, for instance, the hundreds of viewers responding for *Hollywood Squares* or *Entertainment Tonight*.

Science and nature programs also ranked highest on the scale addressing learning. Several documentary and news program formats received high ratings, with science documentaries tied for first with other types of documentaries. Six science documentary programs were among the top 15 programs on the learning scale: the four *Nature* and *NOVA* programs mentioned above, joined by *Wild, Wild World of Animals* and *Smithsonian World*.

A third index addressed emotional impact, asking viewers to respond on a 10-point scale to the statement "This program touched my feelings." While other forms of historical documentaries and plotted dramatic programs led on this index, the six science documentaries mentioned above also registered an emotional effect on their viewers.

One observation to be made from this survey is the high intercorrelation between the indexes. The six science documentaries scored highly on all three scales and, consequently, on a fourth scale combining the learning and emotion scales, given the generic label of "impact." If one accepts that viewers were making distinctions among these three scales, the data suggest that a well-produced science program can impart content, engage viewers emotionally, and be well-liked. This finding has implications for program production as well as future research, in the selection and treatment of topics. For instance, this finding encourages researchers to examine the affective as well as the cognitive domain when studying the effects of programs. It may well be that both domains need to be fully engaged in order to create powerful educational programming, and that the lines between "education" and "entertainment" or "learning" and "drama" may be finer than traditional pedagogy would suggest.

Ironically, one study conducted nearly 25 years ago illustrates an early commitment on the part of National Educational Television (forerunner of the Public Broadcasting Service) to studying audiences for its science programming. NET commissioned an ambitious study on its NSF-funded weekly science series, Spectrum, using a national mail questionnaire sent to high school and college science teachers and a feedback form from small, focused-viewing groups of students enrolled in high school, college, and adult education courses (National Educational Television, 1969). Data were collected for five programs, on topics of astronomy, meterology, seismology, genetics, and an interview show with Harold Urey, Nobel Prize-winning chemist and physicist. This study, while utilizing some relatively simple research methodology, nonetheless provided some interesting and suggestive data.

The science teachers supported the use of public TV to bring science content to a broad public audience and felt the material was accessible to a lay audience. Nearly 85% believed the broadcasts could have the effect of encouraging young people to seek careers in science. While all teachers were sent a copy of a guide for one program, 42% requested additional guides. In a comparison of two programs, the show on seismology and earthquakes was more appealing than a program on genetics. Those samples of adults viewing in groups demonstrated knowledge gains for the program on seismology, while the program on genetics appeared overly technical.

The study also posed a question that continues to be relevant to science productions today: the balance between attracting a voluntary audience and inclusion of "didactic" techniques (e.g., diagrams, summaries of major points) intended to boost comprehension. In addition to program design, the authors also discussed issues of achievable educational objectives for TV programming designed for general audiences and the range of backgrounds and motivations brought by diverse audiences to a single program.

This study is discussed here more to illustrate the types of basic data collected than for any conclusions drawn. (Nor is the appropriateness of a producing organization judging the value of its own programming debated here.) It is unfortunate that more studies such as this one have not been conducted on other science series over the past two decades in order to build a stronger knowledge base for understanding the role of voluntary TV viewing in science learning.

A group of studies, funded by NSF and conducted by Robert Gagne and his colleagues at the Florida State University, has focused on the learning of senior citizens from science programs, specifically *NOVA* (Gagne & Burkman, 1982). In one study, Gagne and his colleagues reedited and shortened a *NOVA* program ("The Insect Alternative") to include voice-over narration (e.g., statements, comments, and questions) intended to support viewers' higher-level understanding of scientific concepts and relationships in the program (e.g., natural selection among insects who have developed immunities to pesticides). This revised sound track provided a form of what cognitive psychologists refer to as "scaffolding," efforts to help learners connect their existing level of understanding to new subject matter and higher levels of understanding. Among a sample of viewers between the ages of 50 and 80, Gagne found that those who saw the elaborated version of the program performed significantly higher on a test of knowledge of natural selection and its social consequences. Both versions received similar appeal ratings.

Gagne conceptualized the "elaborated" version as supporting viewers' "intentional" rather than "incidental" learning from the program, believing that passages in documentaries such as *NOVA* often present lower-level factual knowledge only loosely tied to higher-level conceptual understanding. Acquiring disparate facts can be termed "incidental learning," but does not involve higher-order thinking involved in more intentional, metacognitive learning, or in Gagne's term, "learning control processes."

Gagne also studied senior citizens' "remembering" of main and subordinate ideas in three *NOVA* programs. He found that slightly more than half of the main and subordinate ideas were retained, immediately following viewing as well as two days afterwards. Retention of knowledge was tested through open-ended responses of viewers, rather than closed-ended multiple-choice questions. While Gagne's studies point to directions for future research, his assessment of the state of our research knowledge a decade ago remains true today:

> At the present time, it appears that we know very little about what is learned and remembered from the viewing of TV documentary programs. We have no theoretical basis for the prediction of how much will be learned, what kinds of things will be remembered, and how long they will be remembered. This lack of systematic knowledge applies whether we focus on memory for verbatim recall . . . , the "gist" of stories . . . , or the framework which reflects both traditional and personal aspects, usually called a *schema*.

Another more recent study used the methodology of focus groups, assembled according to specific criteria, to measure viewer interest and learning from a series of *Mr. Wizard* science news inserts (Research Communications, 1987b). Groups were organized in 10 diverse cities across the country. Respondents were screened to include those who viewed local station newscasts in which the inserts were carried. In the groups, they were asked questions of their pre- and post-viewing interest in the topics covered by the inserts, as well as their pre- and post-viewing knowledge of those topics.

On average, the 210 respondents visited science museums about once a year and watched science-related TV programming two to three times per month. According to their own self-reports, they were moderately comfortable with science topics, but somewhat less confident about their science knowledge. The study found a high correlation between topics of interest before actual viewing, and further interest after viewing. Topics of greatest interest included natural dental cement, computerized heart surgery, and problems of low vision, suggesting an audience of older adults. Other topics treated in the inserts, such as computerized mating for cows and the science of tennis rackets, were of lower interest to these groups.

The study also reported a large increase in viewing groups' knowledge of these science topics, based on an eight-item comprehension test:

> Before viewing, participants' average score was 22% correct. Following viewing, participants scored an average of 83%. Consistent with their post-viewing scores, viewers said they felt they learned a little

about a lot. Several said they were surprised at their wrong guesses before viewing and how the simple things they learned amazed them.

The above studies address outcomes related to cognitive and attitudinal change on the part of viewers. Whether exposure to television messages on science can lead to behaviors, such as requesting additional information, reading of books, or going to a science museum, has not been studied with adults, although there is some evidence with children on, for instance, interest in reading of science books.

There is some evidence, from the field of health promotion research, that television and radio messages, when combined with interpersonal support groups, can lead to behavioral changes in lifestyle, diet, and exercise. One of the most frequently cited studies involves a mass media campaign in northern California, conducted by the Stanford Heart Disease Prevention Project (Maccoby and Solomon, 1980). Another more recent example involves an anti-smoking TV advertising campaign in California. Based on a voter initiative, the state of California appropriated $28 million for a TV campaign occurring over 18 months to encourage viewers to stop smoking. From April, 1990 to September, 1991, cigarette consumption in the state fell at the rate of 164 million packs per year, tripling the pre-existing rate.

While both of these examples involved health behaviors, especially coronary disease, they do suggest that TV viewers understood and processed the major messages of those campaigns. Whether scientific literacy can become as important to viewers as their own health and well-being remains to be seen. The inter-relationships between knowledge gained from viewing, affective or emotional responses, and behavioral changes related to science programming remain an important area for future research. For instance, the health-related studies suggest that the media can be used not only to directly communicate scientific information, but also to prompt action, including directing the public to places where further scientific experiences and resources are available (e.g., science centers, public libraries, community groups).

On the Future of the Field

Reconceptualizing "The Field" of Television in Informal Science Education

As the above review of literature indicates, the studies on this topic cannot fairly be aggregated into "a field of research." A critical mass of research has not been conducted, addressing common issues, audiences and outcomes, to serve as a basis for firm conclusions. While a substantial body of research does exist on the effects of television with children (see Comstock and Paik, 1992 for the best recent summary of this work), relatively little of that research relates to educational outcomes with children. Television itself can be viewed as a barrier to science learning, when television programming of non-educational cartoons, game shows, and sitcoms occupies so much of children's time as they grow up.

This review of research raises more questions than it provides answers. There is certainly a plethora of topics for further research. These include the appeal and comprehensibility of science programming for adults; the ways in which viewing of science programming relates to a larger "science habit" for children and adults; the impact of new technological options for access to video material on the viewing styles and learning of viewers; and the ways in which follow-up discussion and further activity can support learning initiated or continued via television.

Before these research questions are studied, a more coherent program of research requires that "the field" be reconceptualized and revitalized. This process must begin by raising the visibility and importance of these questions for communities of television researchers, scientists, science educators, and producers. An example comes from Great Britain, where the importance of informal science education and the role of the media attracted attention at the highest levels of government, science, and education (Ziman, 1992).

By comparison, research on science museum attendance, the design of exhibits, and the response of museum visitors, has grown during the past 15 years (see Bitgood, Serrell, & Thompson, this volume). Much of this research is now reported in a professional journal, *Visitor Behavior*. This growth has been fueled by a professional association, the Association for Science and Technology Centers (ASTC), originally funded by NSF. Science centers are making a greater commitment to conduct research on their visitors, in part because paying customers come through the doors each day. The costs of "rapid prototyping" of exhibits are relatively low (when compared with television production) and visitor reactions can be seen by staff on the museum floor.

In contrast, much of the work of producing science programs for public TV, especially for general adult audiences, has not gone forward with a close connection to viewers. For historical and professional reasons of the craft of television production, producers generally have not sought, nor needed to seek, a detailed understanding of their audiences. (It is ironic that commercial productions for film and television conduct more audience testing than these projects with more educational aims.) The "broadcast mentality" of the television business has tended to erect barriers between producing organizations and their audiences, where global Nielsen ratings suffice for an indication of audience response.

Changing the Paradigm of Evaluation Research: Implications for New Methods and Theories on Informal Learning from Television

In the past decade, the field of evaluation research has begun to acknowledge the shortcomings of strict reliance on traditional research methods and approaches. The assumptions necessary for standard experimental designs—random assignment of treatment and control groups, well-defined independent and dependent variables, validated measures—are typically lacking in most evaluation studies of informal science learning, particularly in studies of television. Experiments or surveys generating quantitative data for statistical analysis often do not capture the richness and complexity of much informal science learning.

The field of evaluation research has been moving steadily towards a more mature and complex view—a new paradigm, if you will—encompassing both theories and methodologies for studying learning from media. This new paradigm has been described in many texts and articles (e.g., Herman, in press; Miles & Huberman, 1984). In general, these authors advocate use of both quantitative and qualitative methods; more clearly defined goals, variables, and measures; sensitivity to the relationships between implementation processes and outcomes; and multiple sources of data and evidence.

Simultaneous with broadening this research to include new methodologies, greater theoretical work on the contribution of television in informal science learning is also needed. This work can begin with an examination of connections to theories being developed in related fields, such as psychology and education. For instance, the field of adult education has undergone some of the same redefinition advocated here. One of the most recent compilations of this work (Merriam & Caffarella, 1991) builds upon the seminal work done by Patricia Cross in the landmark book *Adults as Learners* (Cross, 1981).

During the past decade, researchers in this field have examined the broad range of adult education and learning, ranging from structured, formal instruction to informal and more casual learning opportunities.

These researchers have noted the blurring of distinctions between formal, nonformal (in which television and the media are classified), and informal learning among adults. This field of research has made theoretical progress, describing and debating a theory of "andragogy" for adults, in contrast with theories of "pedagogy" for children. Several assumptions of this theory are relevant to thinking about the nature of adults interested in the range of informal science learning opportunities, including but not limited to television:

- "As a person matures, his or her self-concept moves from that of a dependent personality toward one of a self-directing human being.
- An adult accumulates a growing reservoir of experience, which is a rich resource for learning.
- The readiness of an adult to learn is closely related to the developmental tasks of his or her social role.
- Adults are motivated to learn by internal factors rather than external ones." (Knowles, 1984 in Merriam & Caffarella, 1991)

Studying the nature of "self-directed learning" is an active topic of research in this field. This work promises to make important contributions to our understanding of adults as learners who rely upon a number of sources—television, science and nature centers, national parks, and organizations such as ElderHostel—for informal science learning. Closer communication between the research communities working on these parallel and related issues can help move forward both theory and methodology of research on informal science education. One of the curious gaps in the adult learning literature is the absence of much specific discussion of the role of television, radio, or print media, or community-based institutions, such as public libraries or science centers.

An Increased Commitment to Evaluation Research

Crane (1992) discusses the need for greater commitment on the part of the television-producing community to an understanding of its audience. This commitment should extend to the scientists, educators, and funders who typically collaborate on TV productions in science. As Crane points out, there are many audiences in a national television audience, with diverse educational and cultural backgrounds. These audiences often differ markedly from the backgrounds of, for instance, producers of local science news segments, and process such segments differently than professional producers, who tend to focus on production quality. Developing this commitment to audience-centered program development will require incentives, such as greater funding for evaluation research, and professional training and development opportunities for scientists, educators, and media researchers. While the field has made progress in this direction, especially in the area of children's science and mathematics series, a more comprehensive commitment will be needed to incorporate evaluation on future projects.

A full program of such research should include a cycle of front-end evaluation, or formative evaluation, intended to inform the development of the video product, as well as related print materials or outreach services. Summative evaluation should also be conducted, investigating the cognitive, affective, and behavioral outcomes related to the utilization of the media materials.

The work of these evaluation studies will need to take into account the ways in which the viewing environment has changed in recent years. The viewing environment of the 1990s has become more complex, with new technologies that were not as evident in the 1980s. This environment presents more options for viewers and may mitigate the amount of time and attention viewers give to a particular program. Viewers are now watching television in different ways than was possible before the age of the remote control and access to 50 or more channels. With choices among an increasing number of channels made with a simple flick of the remote control, viewer "loyalty" to any one program has already become problematic.

On the other hand, the technology of cable, home video, and new telecommunications services also make it possible to present a greater amount and variety of science programming than has been possible in the era of broadcast networks. Given changing technologies, it will be important for future research to track viewing behaviors related to these new technological options.

Studies of viewer behavior in this new media environment have begun at the Television Viewing Lab at WGBH-Boston. Simulating the typical living room environment and using sophisticated audience research technology, the Lab has begun to study the viewing habits of adults viewing, for instance, science and nature series such as *Land of the Eagle* and *Mount Everest*. During test sessions, viewers are free to change channels among six programs, have a snack, read, sleep, or go to the restroom. In one study, channel-changing behavior was studied:

> A high percentage of all channel changes occurred during the first few minutes of the viewing sessions. After establishing their program

preferences, viewers made only half as many channel changes during the second 10-minute segment as they did in the first. . . .

More than half of all channel changing could be classified as "grazing." . . . That is, switching to several channels in a single sweep, to sample each for a few seconds. Viewers would frequently return to the channel they were originally watching after making periodic channel scans. . . .

81% of the time was spent watching only one program. . . .

Beautiful scenery, familiar faces, and animals triggered channel-stopping behavior. . . .

One-fourth of all viewing time was spent doing at least one other activity in addition to, or instead of, watching television (Television Viewing Lab, November, 1992).

Although sample sizes are low and the "reality" of the environment is open to debate, this initial research shows the effects of availability of multiple channels and the remote control device on channel-grazing. While some viewers make a commitment to a "dominant program," the study noted that up to half of young adults were "restless viewers" characterized by frequent channel-switching, or were frequently involved in secondary activities, such as reading and eating.

A second study supported the importance of a strong program opening (Television Viewing Lab, February, 1993). Viewers, who tended to be well-educated and cable subscribers, were given an opportunity to view either of two *American Experience* series, "The Iron Road" on the history of the transcontinental railroad, or "Satellite Sky," on the early years of the space race. Viewers in the study became impatient with the slow opening of "Satellite Sky" showing long, predictable scenes of the Earth revolving and rockets lifting off. After 10 minutes of viewing these two programs, and given the opportunity to switch channels among six other programs, viewers exposed to "Satellite Sky" sampled other programming more frequently.

The researchers speculated that the differing structures of the programs may also have affected viewer appeal, comparing the chronological story-telling of "The Iron Road" versus the nonlinear structure of "Satellite Sky," which made viewer comprehension and anticipation more difficult. While it is difficult to directly correlate such issues of program design with audience ratings, contrary to the predictions of the producers themselves, "The Iron Road" attracted higher viewership, a national audience of 5 million people, 32% greater than the 3.8 million who watched "Satellite Sky."

Evaluation efforts that examine the viewing environment should also investigate ways in which viewing by individuals—adults as well as children—can be supported by others in those environments. One striking finding from studies of instructional video indicates that teachers and tutors who reinforce program content with discussions and follow-up activity can amplify the learning gains of their students (Cambre, 1987; Chen, 1991; Johnston, 1987). One of the earliest indications of this effect came from the first summative evaluation of *Sesame Street*, in which a group of preschoolers exhibiting the largest cognitive gains had mothers who were encouraged to watch the program with their children (Ball & Bogatz, 1970). This process of watching programs together and discussing their content has been termed "co-viewing."

In a study relevant to the science learning of adults, the effects of structuring a formal learning environment to include co-viewing, tutoring, and discussion were also seen. The study compared the learning of Stanford graduate engineering students who sat in live lectures, with engineering professionals who viewed videotapes of the same lectures while at their worksites and studying for the same master's degree (Gibbons, Kincheloe, & Down, 1977). The researchers identified the heart of the teaching and learning process as the interaction between learner, materials, and instructor.

> From the standpoint of educational effectiveness, the guideline that is perhaps most frequently overlooked is the one relating to personal interaction, especially where the use of television is concerned. Television . . . does not stop to answer questions; it does not readily permit classroom discussion; it is an inefficient medium for conducting drill; it does not adjust very well to individual differences; and it tends to encourage a passive form of learning.

The researchers developed a method called Tutored Videotape Instruction (TVI), in which small groups of students, together with a tutor, viewed videotapes, stopping frequently to check for questions, problems, and understanding of key points. Using TVI, "students learn best when the videotaped lectures are stopped frequently (for example, every 5 to 10 minutes, for periods of 3 to 5 minutes)." Taking the identical course tests, students who viewed the videotapes under these conditions outperformed those who had the benefit of being on campus and in the presence of the live faculty member. After two quarters, the TVI students had higher grade point averages than those who took the courses on campus. The researchers attribute the success of the TVI students to the opportunities given to students to ask questions and discuss points with each other and their tutor.

The TVI study points to a key element in all viewers' learning from video materials: the availability of knowledgeable others who can enter into an educationally rich dialogue with viewers and enhance, correct, and modify viewers' understandings gained from viewing the video alone. In the world of informal science education, this role can be played by parents; museum "explainers"; community group leaders; and, as this study points out, by other viewers themselves. This role of amplifying viewers' understanding through discussion, dialogue, and human interaction is a crucial one for leveraging the science content contained in the video itself.

Research Over the Longer Term: Beyond "Product-Centered" Evaluation

While a greater commitment to evaluation research focused on specific products of programming, print materials, and viewing environments is needed, a parallel need exists to locate the contribution of media exposure in the longer-term science learning of children and adults. This longitudinal research will need to take advantage of the quantitative, as well as qualitative, tools of social science research, such as field experiments, case studies, and surveys. This research will also need to move beyond documenting short-term "effects" from individual products in narrowly defined settings—TV series viewed at home, museum exhibits, or after-school programs.

In this respect, it is important that the line between formal and informal science education not be too firmly drawn. The distinction is based on the different institutions where learners are found—schools, homes, after-school groups, camps, or museums. Depending on these institutions, differing expectations are set for the role of the learner vis a vis a choice of activities and relationships with others present. The processes of learning, however, should be considered in a more unified fashion across these environments. After all, learners—children or adults—move fluidly between these settings and tend not to categorize their learning into these boxes. Just as reforms in science education encourage an interdisciplinary approach, rather than confining biology to a "biology" class or chemistry to a "chemistry" class, research in this field also needs to focus on the learner learning from a number of different sources over time, rather than short-term studies of individual projects.

This approach acknowledges both the ways in which students learn and the nature of the scientific world. It would be difficult to argue that children learn differently in one context from the other (although their social expectations for behavior may differ in the home versus the school) or that materials offered by informal science learning experiences should change markedly in a school versus a home. There is already evidence of cross-over projects where "informal" materials are used in schools and school-based materials can be used in informal settings. For instance, many museum exhibits travel to schools and television series designed for the home also are used in schools. As cited earlier, Reading Rainbow is the best example of a series originally designed for summer home viewing that is now the most widely used instructional TV series in schools (Corporation for Public Broadcasting, 1992).

One implication of this learner-centered approach is a closer relationship between communities of researchers who have previously worked in relative isolation. Those who conduct research on science centers, community groups, and television projects tend to have different professional training, experience and perspectives. It will be important for these groups of researchers to enter into greater professional exchange if research on informal science education is to move forward.

This field of research also needs to be informed by the advances made in other domains of science education. The work in cognitive psychology on misconceptions in science understanding is one example of theoretical research that should inform studies of informal science learning (see e.g., Gardner, 1992).

Research Over the Lifespan: Bloom's Studies of Talented Young Adults

Several of the studies reviewed, as well as others in science education, describe the important role of the family and parents in encouraging the development of a child's scientific interests. An example of educational research that bears on our understanding of longer-term informal science learning comes from the work of Benjamin Bloom and his colleagues at the University of Illinois. Their studies lay some groundwork for understanding the process of intellectual and personal development, the importance of informal experiences, and the role of key individuals, such as parents, teachers, and mentors, in the learning process.

In their book *Developing Talent in Young People*, Bloom and his colleagues (1985) describe findings from interviews with 120 young men and women, their families, and their teachers, from six fields, including music, athletics, mathematics, and science (research mathematicians and neurologists). Early career achievement criteria were used to select the sample, such as having been awarded the Sloan Foundation Fellowship in mathematics, or a Research Development Award from the National Institutes of Health, or a top ten world ranking in tennis. Those interviewed were asked to reflect upon the individual's personal and intellectual development from the time they were young children, through adolescence, and into their adult years.

The findings are as interesting for the myths they debunk as well as for the importance placed on family influences during the early years. Contrary to the stereotype of brilliant mathematicians and scientists who were child prodigies destined for achievement in those fields, Bloom found that their early years were characterized more by a nurturing family environment, in which intellectual curiosity, hard work, and responsibility were emphasized. As children, these individuals exhibited a range of interests, including being avid readers.

> Talent development is initially viewed by the young child as play and recreational. This is followed by a long sequence of learning activities that involve high standards, much time, and a great deal of hard work. . . . The home environment developed the work ethic and the importance of the individual's doing one's best at all times. . . . Most of the mathematicians felt that their elementary school experience was quite ordinary and considered themselves essentially similar to other students. . . .
>
> The parents valued academic achievement and were models of intellectual behavior. They were typically more highly educated than the average parent, and most had professional occupations. Perhaps the most significant aspect of these early years was the way the parents responded to their children's questions. Questions were treated seriously, and when the parents didn't know the answers, they taught their children how and where to find them. These parents believed that their children were special and shared with them the excitement of discovery. . . . Learning how to learn was more important than what they learned.

Later, as the mathematicians and neurologists progressed through high school, college, and graduate school, teachers and mentors took on increasing importance. The achievements of these individuals were due to a long-term commitment to learning spanning twenty to thirty years. Yet the Bloom study shows how the home environment, the modeling done by parents, and the values transmitted to their children at an early age, were crucially important in setting the stage for later development.

Another theme of the study, echoed throughout much of Bloom's research, is the finding that, in his opinion, 95% of individuals are capable of high achievement, given "favorable learning conditions":

> What any person in the world can learn, almost all persons can learn. . . . At this stage of the research it applies most clearly to the middle 95% of a school population. The middle 95% of school students become very similar in terms of their measured achievement, learning ability, rate of learning and motivation or further learning when provided with favorable learning conditions. (p. 4)

The observations of a *NOVA* producer, who interviewed Westinghouse Science Talent Search winners for a program, also supports Bloom's findings. Linda Harrar (1984) spent several months interviewing these high-performing high school students:

> I learned that the stereotypical whiz kid or genius label didn't fit. . . . they don't seem all that different from their peers. Several are only average students. But somehow, they discovered, early in life, the joys of intellectual challenges. . . . Many winners have had strong support from parents, teachers, or brothers and sisters. Someone along the way helped get them turned on to science. This is one of the things separating them from other bright kids in their classes. . . . At an early age, they develop a keen sense of pleasure in hard work and intense scientific activity.

Both Bloom and Harrar are describing a process of informal science education that applies for all young people, not merely that small percentage that goes on to high-level achievement in the field. The family environment plays the critical role in the awakening of a child's scientific curiosity and the encouragement of his or her interest over the longer term. A child's science learning can be viewed as a garden, tended by parents, grandparents, relatives, teachers and children themselves.

Television programs on science can be viewed as a "genre" of informal educational experience used by families to encourage children's learning. Other genres include science book-reading, trips to science and nature centers, making models, or conducting experiments. In research terms, while television, museums, and after-school activities can be predictors of children's level of science learning, they do not represent the major independent variable, which is more likely the educational background, philosophy, and attitudes of parents. Informal experiences can be seen as key covariates in this relationship. Research should also address the profound equity implications of such processes, which favor families with higher educational levels and stability to focus on their children's educational needs. From a policy point of view, it becomes all the more important to ensure that such experiences are available on the broadest possible basis.

Underlying this view of the importance of the educational values held by families, and how they operate to guide children's learning over the long term, is a view of the importance of time and how time is spent by parents and children. In the critical years of a child's development, from ages two through twelve, many activities compete for that child's time. An extraordinary amount of time is spent by American children with media of many sorts—television, videocassettes, video and computer games. On average, American children watch about four hours of television per day, leading to the often-cited statistic that by the time a child reaches the age of 18, he or she will have spent more time with TV than in the classroom, 18,000 hours with TV compared with 13,000 hours in the classroom. Recent figures show that children with video game players, such as Nintendo or Sega, spend about two hours a day.

When aggregated, these activities, generally not directed towards educational content, consume an inordinate amount of time from children's lives and from the types of educational interactions, dialogue, and activity that can occur between parents and children. Shifting the attention and the ways in which children and adults allocate their time towards educational activities is a major structural issue for the field of informal science education.

Segmenting the Audience: Beyond the Attentive 20%

The sorts of families described by Bloom and his colleagues are well known. They represent families with parents of higher-than-average educational levels, where learning is a critical value in the home. These are the sorts of families who are also active seekers of educational opportunities and activities, including visiting science museums and zoos, reading books, and watching educational TV programs such as those reviewed here.

These families fall into the group of Americans described by Jon Miller as the "science attentive" public (Miller, 1987). One of the critical challenges for the field of informal science education is finding ways to reach the remaining 80% of Americans. As noted earlier, a chief rationale for using television to communicate science is its ubiquity, its presence in 98% of homes. Research on the contribution of this medium to public understanding of science must do a better job of segmenting the "mass audience" and determining what types of viewers are drawn to which types of programs.

A clearer sense of how audiences segment for science programming would then inform thinking on ways to broaden the reach of current and new programming and strategies for reaching underserved groups of women, girls, minorities, and those with lower levels of income and education. There may be other ways of distributing science video materials that may be more effective in reaching these audiences than the conventional ones used in the past.

While a great deal of research remains to be conducted on the issues raised in this chapter, the few studies that exist do point to indicators of the potential audience and impact for science programming. During the next decade, the channels for reaching and distributing such programming, in a variety of forms, will increase. Given the opportunities afforded by these technologies, the need for a research-based approach to the design, distribution, and utilization of video materials for informal learning becomes all the more critical. The challenge to use video technology for the science education of all Americans is as important today as it was when NOVA was developed 20 years ago. What has changed is the urgency to expand the reach of these materials beyond the ranks of the scientifically literate and the well educated. Bloom's research serves to remind us of the potential of a much greater population to be served:

> The central thesis . . . is the potential equality of most human beings for school learning. We believe that the same thesis is likely to apply to all learning, whether in schools or outside of schools. At least, it leads us to speculate that there must be an enormous potential pool of talent available in the United States. It is likely that some combinations of the home, the teachers, the schools, and the society may in large part determine what portions of this potential pool of talent become developed. It is also likely that these same forces may, in part, be responsible for much of the great wastage of human potentiality.

Summary

This chapter reviews the research on the role of television in informal science education. For more than 20 years, television programs on science, nature, and technology have been produced and broadcast on public, commercial, and cable channels. Some of these series represent the highest quality efforts in the history of educational television, such as *NOVA, Nature,* and *The National Geographic Specials.* During these decades, this field also has undergone dramatic changes in technologies of production, distribution, and use by viewers, trends that promise to continue.

However, the cognitive, affective and behavioral impact of these programs remains largely unexamined by systematic research efforts. Literature searching, via both the ERIC database as well as other sources of unpublished research, resulted in less than fifteen studies with data on impact or summative measures. This paucity of research can be traced to a number of underlying factors, such as the "producer-driven" nature of television projects, the absence of funding priorities to conduct summative research, and theoretical and methodological challenges facing such research.

The studies to date point to several outcomes that bear further investigation in future research. The majority of these studies, most conducted with children, suggest that well-produced TV programming can communicate scientific knowledge, especially introducing new topics and basic information to young learners. Such programs can also help to broaden viewers' narrow stereotypes of science and the work of scientists and can motivate an interest in related activities.

This chapter also discusses ways in which this field can be reconceptualized in future programs of research. Recommendations include moving away from a focus on "product-centered evaluation" of single series to a more holistic view of how television, together with other informal and formal sources of education and support, contributes to the science habits of learners of all ages.

Acknowledgements

The author wishes to thank Valerie Crane, Tom Birk, Mike Atkin, Keith Mielke, Mary Budd Rowe and several anonymous reviewers for their comments on the initial scope and a first draft of this chapter. David LeRoy and Craig Reed of PMN TRAC and Sue Bomzer at PBS provided assistance on television ratings data. Karen McClafferty wrote the annotations of research studies. The Cubberley Library at Stanford University, with its ERIC online searching capability, was also of valuable assistance.

References

A. C. Nielsen Television Index (1992, November). New York.

Action for Children's Television (1984). *TV, science, and kids: Teaching our kids to question.* Reading, MA: Addison-Wesley.

Anderson, D. (1992, November). Television, children, and education: Issues for research. Paper presented at the Annenberg School of Communication, University of Pennsylvania. Amherst, MA: University of Massachusetts.

Ball, S., & Bogatz, G. A. (1970). *The first year of Sesame Street: An evaluation*. Princeton, NJ: Educational Testing Service.

Bloom, B. (Ed.) (1985). *Developing talent in young people*. New York: Ballantine.

Boston Consulting Group (1991, February). *Strategy for public television*. Boston.

Cambre, M. (1987). A reappraisal of instructional television. Syracuse, NY: ERIC Clearinghouse on Information Resources, Syracuse University (ISBN # 0-937597-14-7).

Chen, M. (1983). *"Touched by science": An exploratory study of children's learning from the second season of 3-2-1 Contact*. New York: Children's Television Workshop.

Chen, M. (1984). *A review of research on the educational potential of 3-2-1 Contact: A children's TV series on science and technology*. New York: Children's Television Workshop. (ERIC Document Reproduction Service No. ED 265 849)

Chen, M. (1991). *Educational video: What works? A position paper for the Hughes Public Education Project*. Los Angeles: Hughes Galaxy Classroom.

Children's Television Workshop (1990). *3-2-1 Contact research bibliography*. New York.

Children's Television Workshop (1993). *Square One TV: Research history and bibliography*. New York.

Comstock, G., & Paik, H. (1992). *Television and the American child*. New York: Academic.

Corporation for Public Broadcasting (1981a). A comparison of three research methodologies for pilot testing new television programs. Washington, DC.

Corporation for Public Broadcasting (1981b). Assessment of audience feedback systems for research and programming. Washington, DC.

Corporation for Public Broadcasting (1992). *Study of school uses of television and video: Summary report*. Washington, DC.

Crane, V. (1988). The dynamics of informal science learning. Paper presented at the AAAS Annual Meeting, Boston, MA.

Crane, V. (1992). Listening to the audience: Producer-audience communication. In B. V. Lewenstein (Ed.), *When science meets the public* (pp. 21–32). Washington, DC: American Association for the Advancement of Science.

Cross, P. (1981). *Adults as learners: Increasing participation and facilitating learning*. San Francisco: Jossey-Bass.

Exploratorium, The (1981). *Science media conference*. San Francisco.

Federal Coordinating Committee for Science, Engineering, and Technology (1992, August 20–21). *Report on the expert forum on public understanding of science*.

Flagg, B. N. (1990). *Formative evaluation for educational technologies*. Hillsdale, NJ: Erlbaum.

Frenette, M. (1991). Television as a source of informal science learning for pre-adolescents: Design considerations. *Canadian Journal of Educational Communication, 20*(1), 17–35.

Gagne, R., & Burkman, E. (1982). *Promoting science literacy in adults through television. Final report to NSF*. Tallahassee, FL: Florida State University. (ERIC Document Reproduction Service No. ED 229 234)

Gardner, H. (1992). *The unschooled mind: How children think and how schools teach*. New York: Basic.

Gibbons, J. F., Kincheloe, W. R., & Down, K. S. (1977). Tutored videotape instruction: A new use of electronics media in education. *Science, 195*, 1139–1146.

Gore, A. (1984). What have they done to you, Captain Kangaroo? In K. Hays (Ed.), *TV*, *science & kids: Teaching our children to question* (pp. 116–120). Reading, MA: Addison-Wesley.

Gotthelf, C., & Peel, T. (1990). The Children's Television Workshop goes to school. *Educational Technology Research and Development, 38*(4), 25–33.

Hall, E. R., Esty, E. T., & Fisch, S. M. (1990). Television and children's problem-solving behavior: A synopsis of an evaluation of the effects of Square One TV. *Journal of Mathematical Behavior, 9*, 161–174.

Hall, E. R., Fisch, F. M., Esty, E. T., & Debold, E. (1990). *Children's problem-solving behavior and their attitudes towards mathematics: A study of the effects of Square One TV.* New York: Children's Television Workshop.

Harrar, L. (1984). NOVA: The making of teenage scientists. In Action for Children's Television, *TV, science & kids: Teaching children to question.* (pp. 134–139). Reading, MA: Addison-Wesley.

Herman, J. L. (in press). Finding the reality of the promise: Assessing the effects of technology in school reform. In B. Means (Ed.), *Technology in education reform.* San Francisco: Jossey-Bass.

Jerome, F. (1981). Prime-time science and newsstand technology: Is it all just hoopla? *Professional Engineer, 51*(3), 12–14.

Johnston, J. (1987). *Electronic learning: From audiotape to videodisc.* Hillsdale, NJ: Erlbaum.

Johnston, J., & Ettema, J. (1982). *Positive images: Breaking stereotypes with children's television.* Newbury Park, CA: Sage.

Johnston, J., & Luker, R. (1983). *The "Eriksson study": An exploratory study of viewing two weeks of the second season of 3-2-1 Contact.* New York: Children's Television Workshop.

Knowles, M. S. (1984). *The adult learner: A neglected species* (3rd. ed.). Houston: Gulf.

Koran, J. J., & Koran, M. L. (1988). Variables related to learning in informal settings: An overview. Paper presented at the 1987 meeting of the AAAS.

Lewenstein, B. V. (Ed.). (1992). When science meets the public. Proceedings of a workshop organized by the AAAS Committee on Public Understanding of Science and Technology. Washington, DC: American Association for the Advancement of Science.

Lookatch, R. P. (1992, November 11). Passive to active: The use and misuse of video. *Education Week.*

Maccoby, N., & Solomon, D. S. (1981). Heart disease prevention: Community studies. In R. Rice & W. J. Paisley, (Eds.), *Public communication campaigns* (pp. 105–126). Beverly Hills, CA: Sage.

Mariella, R. P. (1976). A scientist views communication with the public. Paper presented at the annual meeting of the AAAS. ED 133 221.

Merriam, S. B., & Caffarella, R. S. (1991). *Learning in adulthood: A comprehensive guide.* San Francisco: Jossey-Bass.

Mielke, K. (1987). Making informal contact. In M. Druger (Ed.), *Yearbook on informal science education.* Washington, DC: National Science Teachers Association.

Mielke, K., & Chen, M. (1983). Formative research for 3-2-1 Contact: Methods and insights. In M. Howe (Ed.), *Learning from television* (pp. 31–55). New York: Academic.

Miles, M. B., & Huberman, A. M. (1984). *Qualitative data analysis.* Sage.

Miller, J. D. (1988). Parental and peer encouragement of formal and informal science education. Paper presented to the 1988 Annual Meeting of the AAAS.

Miller, J. D. (1987). Scientific literacy in the United States. In D. Evered & M. O'Connor (Eds.), *Communicating science to the public.* New York: John Wiley.

National Educational Television (1969). *Science programming and the audiences for public television.* New York. (ERIC Document Reproduction Service No. ED 032 775)

National Science Foundation, Directorate for Education & Human Resources (1991). *Summary of awards: Informal science education, fiscal years 1987–1990.* Washington, DC.

NFO Research, Inc. (1990, August). *Final report: Reading Rainbow study.*

Ormerod, M. B., Rutherford, M., & Wood, C. (1989). Relationships between attitudes to science and television viewing, pupils aged 10 to 13+. *Research in Science and Technological Education, 7*(1), 75–84.

Osborne, R., & Freyberg, P. (1985). *Learning in science: The implications of children's science.* Portsmouth, NH: Heinemann.

Public Broadcasting Service (1987, December 22). *Qualitative ratings report of a national survey conducted by Television Audience Assessment, Inc., April 30–May 13, 1987.* Alexandria, VA.

Public Broadcasting Service (1993, March 30). *1991–1992 national audience handbook.* Alexandria, VA.

Research Communications, Ltd. (1987a). *An exploratory study of 3-2-1 Contact viewership.* Dedham, MA.

Research Communications, Ltd. (1987b). *Research findings for audience evaluation of "How About."* Science Reports. Dedham, MA.

Research Communications, Ltd. (1989, May). *A study of children's informal math learning.* Dedham, MA.

Research Communications, Ltd. (1992, April). *The impact of using the FUTURES series on junior high school students.* Dedham, MA.

Research Communications, Ltd. (1992, November). *Impact of informal learning experience in science, math, and technology: NSF presentation.* Dedham, MA.

RMC Research Corporation (1989). *The impact of Reading Rainbow on libraries.* NH.

Rudy, M. K. (1981). *Sex role stereotypes, interest in science, and responses of 6th-graders to scientists/technologists on the television science series 3-2-1 Contact.* Unpublished doctoral dissertation, New York University. *Dissertation Abstracts, 42*(07), 3095A.

Shapiro, M. A. (1988). *Components of interest in television science stories.* Ithaca, NY: Cornell University, Department of Communication. (ERIC Document Reproduction Service No. 296 400)

Skolnick, J., Langbort, C., & Day, L. (1982). *How to encourage girls in math & science.* Englewood Cliffs, NJ: Prentice-Hall.

SRI International (1988, April). *An approach to assessing initiatives in science education.* Menlo Park, CA.

Television Viewing Lab (1992, November). Couch Potato Chronicles: How people really watch television. Boston: WGBH.

Television Viewing Lab (1993, February). Couch Potato Chronicles: Are there different viewing patterns for high- and low-rated documentaries? Boston: WGBH.

Thomas, L. (1984). Foreword. In Action for Children's Television. (1984). *TV, science, and kids: Teaching our kids to question* (pp. ix–x). Reading, MA: Addison-Wesley.

Tressel, G. (1987). The role of informal learning in science education. Paper presented at the symposium on Science Learning in the Informal Setting, Chicago Academy of Sciences.

Tressel, G. (1990, November). Science on the air: NSF's role. *Physics Today,* 24–32.

Watts, M., & Bentley, D. (1988). Down the tubes: Viewer's opinions of science educational television in the classroom. *School Science Review, 69*(248), 451–459.

Ziman, J. (1992). Not knowing, needing to know, and wanting to know. In B. V. Lewenstein

(Ed.), *When science meets the public* (pp. 13–20). Washington, DC: American Association for the Advancement of Science.

Appendix: Selected List of Television Series on Science and Technology

This list is presented to suggest the range of programming on science, nature, and technology topics broadcast in recent decades. Representative series from commercial and public broadcasting, as well as cable channels, are given. The list has been adapted and updated from an appendix compiled by Paula Rohrdick and found in *TV, Science, & Kids,* published by Addison-Wesley in 1984. Dates signify original premiere of the series and, when known, final year of broadcasts. However, subsequent distribution to cable or broadcast may have extended the life of some of these series. Notation is also made in the case of series currently being carried.

Alive and Well. USA Cable Network, 120 min. 1981. A daily health, nutrition, and exercise series sponsored by Bristol-Myers and produced by DBA Television, Inc., Hollywood, CA.

Animals, Animals, Animals. ABC, 30 min. 1976–1981. A Sunday morning series for children that focused on a different animal each week, produced by ABC News and narrated by Hal Linden.

The Ascent of Man. PBS, 60 min. 1975. A series about how the human race developed and how scientific discoveries influenced the growth of civilization, hosted by Dr. Jacob Bronowski. The series was produced by Time-Life Films, New York, and BBC.

Beakman's World. ABC and The Learning Channel, 30 min. 1992–current. Based on the syndicated newspaper strip "You Can with Beakman," this series for children attempts to demystify science with humor.

The Body Human. CBS, 60 min. and 30 min. 1977. A series of informational specials about aspects of the human body, produced by The Tomorrow Entertainment/Medcom Co., New York.

The Body in Question. PBS, 60 min. 1980. A 13-part weekly series about the human body, written and hosted by Dr. Jonathan Miller. KCET, Los Angeles, and BBC coproduced the programs, which were funded by KCET and Hoffman-LaRoche and are distributed by Films Inc., Wilmette, IL.

The Body Works. Syndicated, 30 min. 1979–1982. A 13-episode series of children's programs dealing with the functions of the human body, hosted by Dr. Timothy Johnson and three 12-year-old assistants. Produced by WCVB, Boston, and syndicated by Metromedia Producers Corp., Boston.

Cable Health Network. Basic cable service, 24 hours daily. 1982–1984. An advertiser-supported cable network with headquarters in New York offering over 20 series on health, science, and lifestyle topics.

The Computer Programme. PBS, 30 min. 1983. A series of 10 weekly programs produced by the BBC that explained computers and their functions. The PBS broadcast was underwritten by Acorn Computer Corp. of Woburn, MA. The series is distributed by Films Inc., Wilmette, IL.

Connections. PBS, 60 min. 1979. A 10-part documentary series on the history of technology, coproduced by the BBC and Time-Life Films, New York. Written and narrated by James Burke and funded by AT&T.

Cosmos. PBS, 60 min. 1980. A 13-part series about astronomy and space exploration written by Adrian Malone and Dr. Carl Sagan and hosted by Sagan. Produced by KCET, Los Angeles, Carl Sagan Productions, and distributed by Films Inc., Wilmette, IL.

Discover: The World of Science. Syndicated and PBS, 60 min. 1982. A series of specials about the latest in scientific development and research, and how scientific progress affects our daily lives. Some were sponsored by Atari, Inc. Produced by the Chedd-Angier Production Co., Boston, in association with *Discover* magazine, and distributed by Young and Rubicam, New York.

Healthbeat. Syndicated, 30 min. 1981. A weekly health magazine series hosted by Dr. Timothy Johnson and produced by Metromedia Producers Corp., Boston.

How About. Syndicated, 80 sec. 1979. A series of 208 science inserts produced by the Mr. Wizard Studio, Canoga Park, CA, and hosted by Don Herbert (Mr. Wizard). Funded by the National Science Foundation and General Motors and distributed by King Features Entertainment, New York.

Innovation. WNET-New York, 30 min. 1983–current. A weekly science and technology series hosted by Jim Hartz that focused on major research advances and implications of scientific breakthroughs. Original funders included Bell Laboratories, AT&T, Johnson & Johnson, and the Robert Wood Johnson Foundation. Innovation specials continue to be produced, such as The Next Generation, profiling a younger generation of scientists.

Launch Box. Nickelodeon, 30 min. Current. A series for children ages 5–10 on the history and science of space exploration, produced by Nickelodeon with NASA and the Astronauts Memorial Foundation.

Life on Earth. PBS, 60 min. 1982. A 13-part series about the evolutionary process from single-cell creatures to humans, hosted by David Attenborough and underwritten by Mobil Corp. Produced by the BBC in association with Warner Bros.

Living and Working in Space. PBS, 60 min. 1992–current. Space professionals ranging from doctors to designers help interest viewers in careers in science and prepare them for a future in which people will live and work in space. Produced by FASE Productions, Los Angeles, with funding from ARCO and the U.S. Department of Energy.

Marie Curie. PBS, 60 min. 1978. A five-part series about the life and the work of the famous woman scientist, starring Jane Lapotaire. Coproduced by the BBC and Time-Life Films, New York and presented by WCET, Cincinnati.

The Machine That Changed the World. PBS, 60 min. 1992–current. A five-part series on the history of the computer, from its earliest origins to present-day applications, such as computer networks. Produced by WGBH-Boston, with funding from Unisys, the Association for Computing Machinery, and the National Science Foundation.

Mr. Wizard. NBC, 30 min. 1951–1965; 1971–1972. An educational program for children, demonstrating scientific experiments, starring Don Herbert as Mr. Wizard.

Mr. Wizard's World. Nickelodeon, 30 min. 1983–current. A magazine-format show aimed at children and adolescents, featuring science experiments, with Don Herbert as Mr. Wizard reprising his role dating back to the 1950s. Produced by Nickelodeon.

National Geographic Specials. CBS/ABC/PBS, 60 min. 1965–1973 (CBS), 1973–1975 (ABC), 1975–current (PBS). A series of documentary specials exploring the beauty and wonder of geography, people, and animals around the world, produced by the National Geographic Society. Currently presented on PBS by WETA-Washington, with funding from Chevron.

Nature. PBS, 60 min. 1982–current. Thirteen weekly programs studying animal behavior, produced by WNET-New York and hosted by George Page.

New! Animal World. The Disney Channel, 30 min. 1983. A daily look at animals around the world, hosted by Bill Burrud and produced by Bill Burrud Productions, Los Angeles.

The New Explorers. PBS, 30 min. 1991–current. Profiles of scientists making the latest discoveries, both in the laboratory and in the field. Hosted by Bill Kurtis.

Next Step. The Discovery Channel and KRON-San Francisco, 30 min. 1992–current. A series on technological breakthroughs in a variety of fields, from medicine to marine science to sports. Produced by KRON for local broadcast and cablecast.

NOVA. PBS, 60 min. 1974–current. A weekly documentary series about science, technology, and medicine. Currently coproduced by WGBH-Boston and the BBC. Over the years, funding has been provided by PBS stations, the National Science Foundations, The Arthur Vining Davis Foundations, Johnson & Johnson, and other corporate and foundation supporters.

Omni: The New Frontier. Syndicated, 30 min. 1981–1982. A magazine-format show that examined how scientific discoveries can affect our future. Produced and syndicated by Omni Productions, New York, in association with *Omni Magazine*.

Powerhouse. PBS, 30 min. 1982. A daily children's and family action/adventure series that provided information about physical and mental health. Funded by a grant from the U.S. Department of Education, the series was produced by the Educational Film Center, Annandale, VA, and was distributed by TeleWorld Inc., New York.

The Scheme of Things. The Disney Channel, 30 min. 1983. A daily, science-and-adventure documentary series for family audiences, hosted by Mark Shaw and produced by Power/Rector Productions, San Francisco.

Science and Technology Week. CNN, 30 min. Current. A review of the week's science news.

Scientific American Frontiers. PBS, 60 min. 1990–current. In this magazine-formatted series of specials, M.I.T. professor Woodie Flowers covers the latest stories on technology, medicine, biology, physics, geology, and chemistry. Produced by Chedd-Angier Productions, Boston, in association with Connecticut Public Television. Funded by GTE Corporation, in association with *Scientific American* magazine.

Schoolhouse Rock. ABC, 3 min. 1973. Animated musical educational segments produced by Newall and Yohe, New York. The Saturday morning series included "Science Rock," "Multiplication Rock," and "Scooter Computer and Mr. Chips."

The Search for Solutions. PBS, 60 min. 1980. A special three-part series about scientists and their work, produced by Playback Associates, New York, hosted and narrated by Stacy Keach. Funded by Phillips Petroleum, which also distributed them for classroom use.

The Secret Life of Machines. The Discovery Channel, 30 min. Current. Originally produced in England, this series explains how everyday machines and appliances work, from the telephone to the VCR to the automobile. Hosted by Tim Hunkin.

The Secret of Life. PBS, 60 min. 1993–current. An 8-part series hosted by David Suzuki examining medical, industrial, and agricultural applications of genetics. Produced by WGBH-Boston with funding from The Upjohn Company.

Slim Goodbody's Top 40 Health Hits. Nickelodeon. 1982. A series of 26 one-minute episodes consisting of songs for young people about good health and nutrition habits, produced by Sheryl Johnston Communications, Chicago, and underwritten by Kraft, Inc. The series has also appeared on CBS's Captain Kangaroo in a longer form.

Spaces. PBS, 30 min. 1984. A six-part weekly science magazine series targeted to minority children, produced by WETA, Washington, D.C., with InterAmerica Research Associates and funded by the U.S. Department of Education and the Alcoa Foundation.

Highlights the accomplishments of minority scientists in order to encourage minority children to consider careers in science.

Square One TV. PBS, 30 min. 1987–current. Daily series for 8- to 12-year-olds intended to make mathematics appealing and to show connections to children's lives. Magazine format includes game shows, music videos, comedy sketches, and a comedy/drama, Mathnet. Produced by the Children's Television Workshop and funded, among others, by the National Science Foundation, Carnegie Corporation of New York, and Intel.

Synthesis. PBS, 60 min. and 30 min. 1977. An occasional series of documentary specials exploring the scientific and technical aspects of public policy issues. Produced by KPBS-San Diego and funded by the National Science Foundation.

3-2-1 Contact. PBS, 30 min. 1980–current. A daily introduction to science and technology in a magazine format for 8- to 12-year-olds, produced by the Children's Television Workshop, New York. Funded by, among others, the National Science Foundation, the U.S. Department of Education, United Technologies, and the Corporation for Public Broadcasting. Currently in syndication to commercial stations.

The Undersea World of Jacques Cousteau. ABC and syndicated, 60 min. 1968–1976 (ABC), 1976– (syndicated). A series of documentary specials about the underwater explorations of Cousteau and his crew of marine biologists, produced by The Cousteau Society and Metromedia Producers Corp., Los Angeles, distributed by Metromedia Producers Corp., Boston.

The Voyage of the Mimi. PBS, 30 min. 1984. A 13-episode weekly adventure series for 8- to 12-year-olds about the study of whales. A second season focused on Mayan civilization, its archaeology, astronomy, and mathematics. Produced by Bank Street College of Education and funded by the U.S. Department of Education. Currently distributed to schools by Wings for Learning, Scotts Valley, CA.

Walter Cronkite's Universe. CBS, 30 min. 1979–1982. A weekly science magazine series, broadcast during the summer, anchored by Walter Cronkite and produced by CBS News. Reports examined topics in science and technology.

The Weather Channel. Basic cable service, 24 hours daily. 1982–current. Daily round-the-clock weather programming, with national, regional, and local forecasts and weather-related features.

What Will They Think of Next? Nickelodeon, 30 min. 1980. A magazine-format series for families about new scientific developments, produced by Science International, Toronto.

Wild, Wild World of Animals. Syndicated, 30 min. 1973-current. A documentary series on animal life narrated by William Conrad. Produced by Time-Life Films and distributed by Ganaway Enterprises, Atlanta.

3

CHAPTER

The Impact of Informal Education on Visitors to Museums

Stephen Bitgood, Ph.D, Beverly Serrell, and Don Thompson, Ph.D.

An exhibition represents a complex set of stimuli, often combining pictures, text material, audio material, models, working demonstrations, lectures, graphs, films, and other methods of presentation. The audience that is attracted to attend the exhibition represents an even more complex set of response potentials. Visitors bring with them a variety of interests, attitudes, educational backgrounds, personal experiences, biases, and intelligence levels, all of these wrapped in larger complexities having to do with age, sex, socioeconomic level, and cultural background. To ask, 'What kind of an impact does an exhibition . . . have on individuals visiting it . . .' is, therefore, to pose an extremely complicated question.

—*Shettel & Schumacher, 1969, p. 59*

Introduction

More than ever before, museums are examining the informal learning impacts of exhibits on visitors. The purpose of these studies is to help shape appropriate methodologies for measuring impact, to better understand the nature of the learners and the impact itself, and to improve the quality of museum visitors' experiences.

Many historic and current studies have contributed to our understanding of visitors, exhibits, the total museum experience, and the short and long-term impacts of visiting a museum. There is still much to be learned. Professional organizations have grown and numerous conferences and publications are available to inform the profession and help develop theories and practical advice. The area of visitor studies is no longer a minor influence in the decisions and directions museums will make and take in the future.

The increase in the number of opportunities for informal learning in science museums, zoos, and aquariums (as well as other types of museums) for visitors has been accompanied by shifts in the goals and administrative structures of museums, as well as the creation of many new institutions and renovations and expansions of existing ones. A new commitment by museums to focus on audience needs has been encouraged by desires to increase effectiveness of exhibits and to reach underserved populations within the museum's community.

Various philosophical and methodological arguments that have slowed progress or inhibited change in the past are being discussed in more integrated and positive ways now. This can lead to more rapid improvement in exhibit development procedures and the integration of evaluation techniques to enhance visitors' opportunities to learn about science in the informal museum environment.

Although most museums are currently enjoying a secure and popular place in modern culture, their role and accountability as educational institutions have been questioned. Methods to capture and document the efficacy of that important role have reached a critical mass, and with more shared data across institutions, theories will emerge to guide the continuing development and refinement of the techniques of education that work best in museum settings. The future is bright for museums as they adapt to and address the challenges of teaching science in a pluralistic and increasingly diverse society.

Throughout this chapter, the reader will be directed to sources in the literature for more in-depth treatments of issues and topics under discussion. The point of view to be presented here comes from authors whose main concerns are the planning, design and evaluation of exhibitions and who seek to optimize the visitor-exhibit interaction. For a more detailed history of visitor studies, see Shettel, 1989 and Bitgood & Loomis, 1993. For a short discussion of visitor studies from a more theoretical and psychological viewpoint, see Uzzell, 1993. For a review of the integration of the concept of visitors in exhibitions, see Schiele, 1992.

The objective of this chapter is to provide a brief sampling from the visitor studies literature (mainly from the U.S.A.), preceded by the conditions, limitations and opportunities that museums offer, and followed by the issues that have influenced those studies and what the next steps might be.

Museums as Sites for Informal Learning

Science museums, zoos, and aquariums are places where informal learning can occur naturally and logically, creating an exemplary model for other types of museums.

Several special characteristics afford learning in museums both advantages and limitations. First, in the informal exhibit environment, affective and cognitive learning experiences are fused, not separately structured activities or objectives. Second, and in concert with the first, education and recreation are not dichotomous. Both are happening at once. Third, exposure time—i.e., time spent by visitors with individual exhibit elements—is usually brief and often unstructured. "Time on task" is measured in seconds and minutes, not hours at a time. Thus, learning experiences tend to be brief, fun, episodic, self-selected and self-controlled. Visitors are not under any obligation to learn anything.

The above aspects apply to most other types of museums where there are exhibits with a didactic or narrative component or objective. The conditions for informal learning and the techniques for enhancing learning opportunities are similar and the problems of interpretation and communication are the same. The specifics of what is learned, rather than how, vary in different types of museums. Besides, in any discussion about science exhibits, it's hard to separate out just science museums—there are many combinations, such as, science and art; natural science and history; history and industry; and, zoo and museum.

There are several other general aspects of what the learner does in an informal setting that are particularly appropriate to museums; in fact, museums may have the most creative opportunities for their expression. These activities include:

a. Making quick connections between what is personally known and something new, resulting in new associations and relationships;

b. Having an authentic experience: seeing the real stuff (e.g., objects, artifacts, animals), or experiencing the actual phenomena, or having access to the accurate, simulated device;

c. Having experiences that involve naming, identification, observation, imagination, fantasy, imitation and role playing, cooperation, demonstrations and discovery;

d. Being able to covet objects (guiltlessly);
e. Having no limits, tests, or lectures.

In the November 1993 issue of the magazine *Discover,* in "Ten Great Science Museums," ten authors of different interests, gender and background describe their informal and sometimes seminal museum experiences, using terms that reflect the above list, such as "discovery," "passion," "relevance," "magic," "awe," and "reverence."

Learners who engage in these activities are not limited to any particular age group, gender, cognitive style, income level, or cultural groups, etc., although when and where they become engaged will be influenced by these variables. While each person's museum experience is unique, there are many shared human reactions and response patterns (e.g., attention, memory, reasoning, feelings, and motor skills) that can be systematically studied and described. While there is no such thing as "the average visitor," averages (e.g., of time spent or the number of exhibits stopped at) can suggest trends and allow comparisons to be made between different samples.

To understand the eclectic nature of how learning occurs in museum settings requires borrowing and adapting theories and techniques from many other fields, such as education, psychology, and mass communication. Museums have learned much and still have a lot to learn, but ultimately no one can tell them more about themselves than they—the practitioners and the visitors—can.

Those science people, they have it easy. They've got the scientific facts to work with. They know. There's no debate.

(A history museum staff person)

Those history people, they have it easy. They've got the historical facts. There's no question—it happened.

(A science museum staff person)

A Plethora of Opportunities

The intent to provide informal education opportunities and activities in museums for broad-based community involvement has never been greater.

Starting in the 1970s, the number of institutions—science museums, children's museums, aquariums—has expanded, and the trend has not slowed down. In 1992, more than 25 new institutions were in the planning stages. Museums are popular, drawing high attendance figures, and they have provided catalysts for inner city revitalization.

Museums are conventionally regarded as being educational, by themselves and by their communities. Internally, the museum's educational role is often written explicitly in its mission statement. Externally, the community's conventional rhetoric identifies museums as cultural icons, where learning and entertainment co-exist. Consider the following hype for a new science museum exhibit in a travel magazine: "Learn about motion and equilibrium as you spin in the Gyro Chair. Ride a bike to generate electricity to illuminate 15 light bulbs. Strange illusions make you appear to fly, multiply, and switch faces with a friend. It's so much fun you won't even know that you're learning until you've already grasped basic scientific principles of vision, sound, energy, and gravity."

Large numbers of programs and educational exhibits (permanent, traveling and temporary), outreach and satellite programs and exhibits are made available by museums that make exhibits for rent to other museums. Collaboration enables more cost saving and sharing. Many of these traveling exhibits have an emphasis on active and popular characteristics—interactive and participatory elements, theater, demonstrations, and high technology.

More museum staff than ever before have special training to do their jobs. With more expertise on staff—education, design, media, museology, business management—comes new approaches to how museums function internally.

The traditional approach to exhibit design was to have a subject/content person research and write copy and then hand it off to a designer to lay out the exhibit plan and select the media. After the exhibition was opened, the education department would develop and run programs for it. Many museums have made a shift from this linear approach to the planning and design of exhibits as a team approach. The team approach often includes emphasis on an "audience advocate," and integrated educational programming occurs earlier in the process.

Along with the greater availability of educational exhibits has come more emphasis on accountability and evaluation. If an exhibit is to be traveled and shared with several institutions, the borrowers or renters feel more confident in the show if it has gone through some evaluation procedures either during its development or after its first venue. In many cases, evaluation has been a requirement by the consortium or granting agency that helped fund the exhibit.

A stated role of science museums is to show phenomena and processes, and inform visitors about scientific facts and issues, "authoritative information in a setting where people can keep abreast of new developments in science" with the intent of having visitors gain "enough

facts to enable people to make intelligent decisions on their own"
(Kimche, 1978). Idealized as that statement may be, there is little question about the popularity and presentation of opportunities and the
potential to have impact. *Actual* impact is less understood.

> When I was a kid, I thought it was called The Museum of Science and
> <u>Interesting</u>.

                                   (an adult museum-goer from Chicago)

New Dedication to Audience Needs

**Underlying these efforts has been the power shift within some muse-
ums from curatorial to educational concerns, that is, from primarily
object-based interests to primarily audience-based interests.**
Many museums have undergone recent organizational restructuring
of their public programs and exhibition departments. At the Field Museum of Natural History, where many halls of aging exhibits were to go
through a period of renovation, Mike Spock was hired in 1986 to direct
the public programs department that included the education and exhibition departments (design and production) and established a new division
of exhibit project directors. The scientific research functions of the Field
Museum were made more distinct, releasing curators from traditional
but irregular roles in charge of exhibit development, making them
responsible for content accuracy and object conservation but not in
charge of the exhibit process. At the Natural History Museum in London,
a similar change had occurred earlier, in 1972 when their "New Exhibition Scheme" was launched. In both institutions, and others that have
made similar changes, the restructuring was not always welcomed by the
scientific research curators, some who felt that exhibition content accuracy and integrity would be sacrificed or compromised.

The intent of the changes from having curator-controlled exhibits to
exhibit developer-led projects (by people who were sometimes general-
ists, sometimes subject specialists) was to develop exhibits with more
popular appeal and effectiveness and move away from the elitist, exclusive, rarefied atmosphere that many collections-based exhibits seemed to
have.

Many science-technology centers do not have collections, and therefore do not have research curators. Staff requirements in these "museums"
often combine subject matter expertise and skills in public education. At
New York Hall of Science, in Corona, New York, Alan J. Friedman started
in 1984 with an unusual organization where separate departments based

on subject/topics (e.g., biology) would each have their own education and exhibition components. After a few years, a more traditional approach was adopted with conventional departments to consolidate expertise and efforts and avoid duplication of job tasks, but still retain the advantages of a project team approach (see Friedman, A. J., 1991, for more about science museum management style).

This shift in process and attention to intended audience is not just an isolated, unique occurrence for museums. A broader, cultural shift underlies it: pluralism, systems thinking, and multiculturalism have become more important, and fundamental philosophies have changed as well. No single group is seen as having exclusive access to knowledge in the postmodern mentality. It's more a matter of interpretation, narrative, and storytelling, and the story depends on the teller. (For more discussion of this issue as it is acted out in museums, see "From Knowledge to Narrative," Roberts, 1993.)

Although science museums seem obviously audience-based with their "fun things for people to do with science," they sometimes still are guilty of having an elitist and condescending attitude toward visitors, such as, "We can't really expect the visitors to understand the scientific principles anyway." This assumption keeps science on a pedestal and inhibits the creation of exhibits where visitors get to help define the content. When the museum's intent (e.g., the scope of the communication objective of a science exhibit element) has realistic overlap with the audience's behavior, attitude and expectations, learning is more likely.

In addition to the changes in society (political, economic, philosophical) that helped move the museum's internal restructuring to a more audience-centered approach, there has been more practical support in the form of professional development for museum staff. More than 500 participants from every kind of museum participated in the Kellogg Project at Field Museum, six years (1982–1988) of workshops "empowering educators" and promoting the team approach. The publication of *Open Conversations* (Blackmon, *et al.*, 1988) was a synthesis of those six years. It gave museum staff the tools to conduct similar workshops themselves to improve their understanding of the museum's educational role, the visitor's experience and communication through exhibitions.

The American Association of Museums called for education to be a primary activity in every type of museum in *Museums for a New Century* (Bloom, *et al.*, 1984). Four of the Commission's 16 recommendations were directly related to education, and in *Excellence and Equity: Education and the Public Dimension of Museums* (AAM, 1992) there was even a stronger call for commitment to, inclusivity of and leadership for the educational role of museums.

These changes are all helping museums to become less formal, and as such, to become a better match for their very informal audiences. Readers of this chapter might want to investigate more about the particular museums that were significant models (in the U.S.A.) for these trends over the years: Milwaukee Public Museum (50s and 60s) for an early dedication and attention to the public's interests and needs and visitor research (De Borhegyi, 1963); Boston Children's Museum (60s and 70s) for examples of visitor-oriented exhibits; The Exploratorium (70s and 80s) for developing hands-on participatory science exhibits that have been shared and copied throughout the world (Oppenheimer, 1986); and the Field Museum of Natural History (80s and 90s) for an institutional-ized commitment to the museum as an informal learning environment and to visitor research.

When all is said and done, the history of visitors to the exhibition is basically the history of what we expect of them and the means used to induce them to act accordingly.

—B. Schiele, 1992

Professional Organizations Offer Support

Many professional organizations have focused on the nature of infor-mal education in museums and support dialogue, meetings, and publi-cations for the dissemination of findings related to the theory and practice of visitor studies.

In 1986, Steve Bitgood, at Jacksonville State University, Alabama, began publishing a quarterly newsletter called *Visitor Behavior: A news-letter for exhibition-type facilities,* dedicated to the study of visitors in informal learning settings. Bitgood's publication unknowingly scooped long-time intentions by Chandler Screven (psychology professor at Uni-versity of Wisconsin and museum studies guru) to establish a visitor studies organization, namely the International Laboratory for Visitor Studies (ILVS), which he finally launched two years later with *ILVS Review: A Journal of Visitor Behavior.* Bitgood and Screven allied in an informal, cooperative relationship, so that people who received the review also received the newsletter.

Years before (in 1977) at the American Association of Museums (AAM), members interested in visitor studies had formed the ad-hoc committee on evaluation, which became an official Standing Professional Committee in 1993. Now called CARE (Committee on Audience Research and Evaluation), the committee hosts a poster session of short papers on recent research at yearly AAM conferences and publishes the papers in *Current Trends in Audience Research & Evaluation*, available each year. CARE has recently published its "professional standards," a document that sets guidelines for both people interested in doing visitor studies and supporters (e.g., museum managers) of the activities of visitor studies practitioners.

Visitor studies conferences have been held from 1988 to 1991 and the proceedings were published by the Center for Social Design in Jacksonville (headed by Steve Bitgood and Arlene Benefield). In 1992 the Visitor Studies Association (VSA) was officially formed by Bitgood, Benefield and a core of roughly a dozen others who had been active in the AAM's evaluation committee. Around 300 people have attended conferences each year in 1991–1993.

Many of the subscribers to *Visitor Behavior* and *ILVS Review* and members of CARE and VSA are the same people. The current overlaps in objectives, duplications of effort by personnel (as editors, committee members, etc.) and stretched financial resources (separate dues) may lead to a merger, or to more distinctly specialized missions (e.g., providing practical information vs. a peer-reviewed journal of academic standards).

Examples of other professional organizations and publications with relevance to informal learning impact studies in museums follow:

1. The Association of Science-Technology Centers (ASTC) published a series of papers on "What Research Says About Learning in Science Museums" (ASTC, 1990 and 1993) covering such topics as "How Cognitive Scientists View Science Learning," "Holding Power: To Choose Time Is to Save Time," and "What Have We Discovered About Discovery Rooms?"

2. *Curator* magazine, whose new editor, Sam Taylor, has taken a new direction in encouraging and publishing more visitor study-related articles.

3. The Education Committee (EdCom) is one of the oldest and biggest Standing Professional Committees within AAM, with many members interested in visitor studies.

4. Members of the National Association of Museum Exhibition (NAME) receive the quarterly *Exhibitionist* that contains articles related to visitor studies.

5. American Zoo and Aquarium Association (AZA) holds regular annual conferences and publishes the proceedings that feature visitor studies-related information dating back to the 1970s. Zoos and aquariums are particularly interested in impact as it relates to visitors' attitudes about environmental issues and wildlife conservation.

6. In Washington, D.C., the Museum Education Roundtable (MER) has meetings and publishes the *Journal of Museum Education,* with museum audiences and impact a frequent topic.

Why are so many people interested in visitor studies? Some people find it fun, worthwhile, moral, democratic, and creative, although not particularly lucrative. One side effect of doing visitor studies is that it can promote an understanding of human nature: after watching and talking to thousands of people in museums, one witnesses the goodness of human nature, how hard visitors really do try to understand exhibits, how much the audience appreciates exhibits and museums, and how there is no such thing as the so-called lowest common denominator among museum visitors.

Other museum people become interested in visitor studies because they have been told they have to do it by administrators or funders. Their first foray into it is often a phone call to an evaluation consultant, instead of a trip to the library to read about it.

Hello? Your name was given to me. . . . We want to do an evaluation. How much will it cost?

(Museum professional with a naive notion)

Research and Evaluation on the Nature of Impact

Research on the nature of the impact of informal education in museums has reached a critical mass—many trends, issues, and findings have been verified.

There is a more than 90-year history of studying impacts on visitors through observation, testing, interviewing and surveying. Practitioners, academics, researchers and students have looked at who visitors are, what social and psychological constructs they bring with them, what they do in museums and exhibits, and what they recall about their visits.

Some studies have been done to find out if an exhibit is "working." Those studies are often driven by the desire to improve the functioning of exhibits. A single element within an exhibit, such as an interactive computer, or an entire exhibit may be the subject of a study. Other studies are driven by the desire to understand more about the audience, to learn about their characteristics and needs. These studies may be looking for trends among visitors, perhaps for marketing purposes, because by knowing who audiences are and what they like, museums might know how to better attract them.

Whether altruistic—in the name of pure research—market-driven by economics, or pragmatic attempts to optimize an exhibit, studies of museum visitors have ranged widely in terms of purpose, scope (e.g., number of subjects, cost of the study, years spent doing the study) and usefulness of the findings, as well as amount of competence and types of bias the researchers have.

Examples of Experimental and Single-Variable Studies

The most thorough list of research and evaluation conducted over the years throughout the U.S. and around the world can be found in the *Visitor Studies Bibliography and Abstracts* (Screven, 1993), now in its third edition. It is an invaluable resource for finding information about impacts of informal learning in museums.

In the bibliography under "Experimental Studies," there are many reports that used experimental design conditions and show a positive impact on visitor learning from museum exhibits. Several variables that have been studied repeatedly over the years by many different researchers are orientation aids, touring aids, time as a dependent variable, and participatory or interactive exhibit elements.

- Orientation aids which help visitors (both children and adults) prepare for the visit (both before arriving and on site), such as instructions and floor plans on signs or hand-outs, can increase efficiency of use, time spent, attention and recall. Using an interactive computer that reinforces the exhibit's theme and content before or during a visit can help sensitize visitors and increase the number of stops they make and time spent. Even a pretest as part of an evaluation study can serve as an orientation for visitors to the exhibit, enhancing specific learning from it. (For details on the individual studies, see Screven, 1973; Gottesdiener & Boyer, 1992; Goldberg, 1933; Bloomberg, 1929; Hilke, 1988; and Hayward & Brydon-Miller, 1984; Gennaro, *et al.*, 1982; Cohen, *et al.*, 1977.)
- Self-guiding devices used to augment a visit, such as a worksheet, brochure, audiotour, or guidebook have been shown to increase informal science learning for children and adults. (See DeWaard, *et*

al., 1974; Robinson, 1928; Korn, 1988; McManus, 1985; Screven, 1975; Reque & Wilson, 1979.)

- In many cases, the amount of time spent in an exhibit is positively related to the amount of learning. (See Abler, 1968; Barnard, *et al.*, 1980; Falk, 1983, 1993; Nedzel, 1952; Porter, 1938; Peart, 1984.)
- Participatory devices in exhibits often attract more attention and time from children and adults than do static exhibits, and learning gains can result. (See Brooks & Vernon, 1956; Eason & Linn, 1976; Koran, *et al.*, 1986, Borun, 1977.) Long-term impacts were documented from "The Launch Pad," an interactive exhibit at the Science Museum in London (Stevenson, 1992).

Most of the above studies focused on environmental variables that can facilitate impact. Other reports look at impact as it is influenced by the characteristics of the learner, such as personal perspective (Anderson, *et al.*, 1983), cognitive development (Boram, 1991), thinking skills (Greenglass, 1986), and social groups (McManus, 1987) to name only a few. Family groups and gender have been studied in many exhibit settings. Broad conclusions are more difficult to draw from these, however, because there are many confounding variables, not only between visitor variables (such as age, education level), but between visitor variables and exhibit variables. For example, Rosenfeld (1979) and Diamond (1986) found that children led family groups to exhibits. Depending on the type of museum, families may make up 20% to 80% of the audience. Because exhibits should function well regardless of who gets there first, it makes sense to put efforts of studying impact into those variables that can be manipulated easily and changed if improvements warrant them. Getting to know more about audiences is important, but museums have more control over and responsibility for choosing exhibit variables than they have over their self-selecting visitors.

Experimental studies in museums often have used school groups as the test subjects not only because of the importance of that segment of the museum audience, but also because of the familiarity of organizing the experimental design in a way similar to other educational impact studies. (Several of the examples cited above and elsewhere used school children as subjects.) These studies often have important implications for non-school group audiences—the main focus for this chapter.

Looking again at the *Visitor Studies Bibliography*, one finds there are fewer studies representing those which are traditionally experimental, that is, where cause and effect are sought, control groups are used and single variables are studied one at a time. Rather, many visitor studies are descriptive, theoretical, or evaluative, or a combination of these.

The number of non-experimental, empirical, and evaluative studies done on the impact of a single exhibit element is constantly growing. Labels and computers are among the elements singled out for more intensive study by many different researchers, besides the elements listed above (orientation, participatory elements, and self-guiding devices).

- Labels have been the focus of attention in all types of museums. The effects of label placement, lighting, length (number of words), type style, paragraph length, use of catchy phrases or questions have been theorized about and tested. Often, however, a lack of control of variables in these studies inhibits the generalizability of their findings. One conclusion that has been drawn from multiple sources is: less is more. That is, shorter labels attract and hold more visitors' attention longer to read more (Bitgood, *et al.,* 1989; Bitgood, 1990b; Borun & Miller, 1980; McManus, 1989a; Screven, 1992; Serrell, 1981; Wolf & Smith, 1993). There is evidence that labels written with interrogative and imperative formats are more likely to communicate (Screven, 1974, 1986; Hirshi and Screven, 1988; McManus, 1990), but the kinds of questions should be carefully considered. Improvements in label texts can lead to overall improvements in the learning impacts (Falk & Dierking, 1992).
- The impact of computers on visitor learning and how computers influence visitors' use of the whole exhibit has been studied by a few researchers, but not as extensively as the cost of these elements might warrant. Visitor behavior has shown that computers are attractive and popular exhibit elements, especially for younger visitors. (For descriptive studies discussing the use and impacts of computers in exhibits, see Hilke, 1989; Hilke, *et al.,* 1988; Screven, 1990a; Serrell & Raphling, 1992; and Flagg, 1991.)

More exhibits that immerse visitors in a total environment for experiential learning are being built at zoos and aquariums, as well as in natural history museums. The single variable of sound was studied by researchers Ogden, Lindburg, and Maple (1993) at the "Gorilla Tropics" exhibition at the San Diego Zoo. Visitors who were exposed to the sounds (controlled by 27 sensors detecting visitor movements and controlling more than 100 speakers) were compared with visitors who were present when the sounds were turned off. The study suggests that the addition of appropriate sounds to a naturalistic exhibition can have an important impact on cognitive and affective learning. For example, two of their findings were:

—When the sounds were on, almost twice the number of visitors said they had learned something in the exhibit than when the sounds were off.

—When asked if the exhibition had influenced their emotions/ feelings toward the animals or their habitat, significantly more visitors reported positive influences when the sounds were on than when off.

These results are consistent with other studies that suggest the experience is improved when the exhibition makes the visitor feel immersed in the exhibition environment (e.g., Bitgood, 1990a).

Measuring affect is not new, but more recent interest in it is evident. For methods that emphasize affective aspects of informal learning, visitor-centered learning studies and some innovative ways of collecting data, see Roberts (1990), Raphling & Serrell (1993), Silverman (1990), and Edington, *et al.,* (1993).

Studying the Studies

What do we know about the field of visitor studies and the intellectual context of trying to study impact? The answer is more than we know about the impact itself. Context has had important implications for the issues of studying visitors and for finding out why we don't know more now than we might. Knowing something about the trends and history of visitor studies becomes an impact study in itself. To demonstrate, the following comments chronicle a few of the "fits and starts" of early attempts to study impacts. (For more thorough treatments, see Bitgood and Loomis, 1993, and Screven, 1993.)

After a beginning of many smaller studies done from the 1900s through the 1930s, there were several big-scale experimental studies done between 1952 and 1976, using large sample sizes, multiple techniques of gathering data, and complex data analysis. Some of these studies had results that could be interpreted as being less than flattering for the institutions where they were conducted, and thus, there was little institutional commitment for the dissemination of the findings. Also, some of these comprehensive reports were long, technical, or not published in readily available museum or educational literature, and therefore, inaccessible to many readers.

- Nedzel (1952) studied the motivation and education of the general public in an exhibit on magnetism at Chicago's Museum of Science and Industry. Time-lapse photography was used to record and track visitor movement and behavior in the exhibit, and post-tests measured cognitive gain. There were no learning gains for the normal visit conditions group, although groups given a pretest, special signage or a lecture tour did show positive learning impact. One of her conclusions was that "objective appraisal is essential if the museum is to make serious claim to . . . an educative or culturally significant role." Nedzel identified many of the factors that could contribute to a more successful learning experience for visitors, and she suggested more attention be given to studying "indirect effects" on visitors. As an unpublished doctoral thesis, this study—a model of thoroughness—never reached many viewers, and Nedzel did not pursue a career in museums.

- Taylor (1963) reported on an extensive study of a Seattle World's Fair exhibit on science and concluded, like others, that the communication of ideas and principles by means of exhibits is very difficult. Weiss and Boutourline (1963, discussed below) under contract with a large industrial firm also studied visitors to a fair site, as did Shettel (1967). These large studies have not been widely circulated or used by museum practitioners, probably due to the bulk, technical approach and somewhat obscure publishers of these reports.
- Lakota (1975) used a multi-method approach to studying visitors at the National Museum of Natural History. The work was commissioned to provide an empirical framework for planning future exhibits at the Smithsonian Institution, but it was not made widely available to museum professionals. Even if it had been well circulated, many readers would have found the technical language and statistical analysis difficult to follow. Far more readable were Lakota's later recommendations in "Good Exhibits on Purpose: Techniques to Improve Exhibit Effectiveness" (1976).
- Shettel (1976) evaluated "Man in His Environment" at the Field Museum in Chicago. Attitudes and knowledge were measured by multiple-choice test items; a post-test interview with open-ended questions (e.g., "Tell me what you think the basic message of the exhibit was?"); a questionnaire in which visitors were given the prompt: "I think of nature as . . ." with bipolar descriptors (e.g., living-dead, complex-simple, not valuable-valuable); and a 16-question attitude questionnaire related to man and nature. In terms of knowledge acquisition, only one educational level (those who completed high school) showed a significant difference between pre- and post-test for the multiple-choice test items, but most showed small, positive changes. In terms of attitudes toward nature,

little difference between pretest and post test was observed, probably because the pretest responses were so positive there was little
opportunity for change. Shettel made numerous recommendations
in his evaluation of visitor response to the exhibit, but the findings,
while useful and insightful, were apparently not incorporated readily
into subsequent exhibition development processes. In the years
following this study, he often quoted it as an example of the failure
to institutionalize evaluation.

One short study that summarized some of the earlier and seminal
works by Robinson (1930) and Melton (1935), that received attention,
was frequently quoted and subsequently has had many of its ideas
confirmed, was published in 1963 in the popular professional journal
Museum News by Weiss and Boutourline, titled "The Communication
Value of Exhibits." In the article, they made the following generaliza-
tions:
   —Large, vivid exhibit elements become "landmarks" in exhibitions
   —Visitors' paths tend to be prominent between landmarks
   —The most direct route from the exhibit entrance to the exit is most
used
   —Visitors rarely zig-zag between exhibits that face each other, and
therefore, facing exhibits compete with each other
   —Complex, difficult concept exhibits get visitors' awe and respect but
not comprehension
   —Different styles of exhibits elicit different ways of responding, and
adults and children respond in different ways
   —Exhibits function best when they relate to visitors' prior interests
   —"Effective communication of knowledge, as distinct from creation
of an experience, is a very difficult task within the museum situation."
(p. 26).
   In their admittedly small study (in an art museum), they pinpointed
many of the problems and issues that would continue to perplex and
challenge museum practitioners in all types of museums for the next
thirty years.

Recent Exemplary Studies

In the last ten years, three books detailing multi-method studies, positive
results on informal science learning and representing good models,
broader definitions of learning, with easy to share findings are: *Families,
Frogs, and Fun*, by Judith White and Sharon Barry (1984); *Try It!*, edited
by Sam Taylor (1991); and *New Dimensions for Traditional Dioramas*, by

Betty Davidson (1991). All three of these publications contain exemplary processes and products for increasing exhibit learning impacts.

- *Families, Frogs, and Fun* is White and Barry's report on the impact of HERPlab, a zoo learning center designed to teach families. Most of the results of this report related to data collected at the zoo on the immediate impacts on users. However, one aspect of this study was a small, long-term follow-up phone survey of people two to three months after their visit. These investigations found that 19 of 21 people interviewed reported that they did something as a result of their visit. This included talking about the visit afterward, telling other people about the visit, making a return visit, reading books on animals, purchasing microscopes for children, and applying what they learned (e.g., teacher using experience for classroom, family looking for territorial behavior in their cats and on TV programs). The success of this project has led to the development of more in-exhibit, hands-on interactives related to reptiles that have been tested and shared with several other zoos.
- In *Try It!*, eight case studies where formative evaluation was used to gauge impact and improve the effectiveness of exhibits before the exhibits were finished are described. This book is not about traditional research; rather, it features examples of using practical, small sample, quick feedback evaluations for trying out an exhibit while it is still in the development stages. The cost-benefit ratio of such efforts is high, as relatively easy modifications are shown to dramatically improve visitors' understanding of labels and correct operation of exhibit devices. The case studies show that during formative evaluation, crucial visitor feedback informed exhibit developer's decisions, resulting in better displays and more successful interactives. In addition to the case studies, *Try It!* is about the hows and whys of evaluation and discusses the benefits of evaluation to museum management as well as to visitors.
- *New Dimensions for Traditional Dioramas* resulted from renovations to exhibits at the Museum of Science in Boston designed to accommodate special needs visitors, such as people with visual, motor, or auditory limitations. When exhibits were made more accessible to disabled visitors, benefits were available to all visitors. Pre- and post-evaluations showed notable increases in the amount of time visitors spent, the thoroughness with which they investigated the exhibit area, and visitors' knowledge of the exhibit themes. The process that the exhibit developers went through to plan and imple-

ment the changes, as well as the "products" (e.g., kinds of changes—physical structures, space allotment, tactile and multisensory experiences, legibility issues) can serve as good models for other institutions.

Institutional Commitment to Studying Impact

At the Natural History Museum in London, the trials and tribulations of doing visitor surveys and measuring impact, using the findings and being criticized for taking a visitor-oriented approach, is a story unto itself. Roger Miles, a noted paleontologist, was made Head of the Public Services Department in 1975. Under his leadership, assisted by Mick Alt, Steve Griggs and others, the then-called British Museum (Natural History) became the first museum to adopt an internal, systematic approach to visitor evaluation. First, they tried to apply working assumptions about effective design of exhibitions in "Human Biology," and bore the subsequent critical reactions from peers (curators and academics) and favorable responses from visitors. Countering the criticism, Alt & Shaw (1984) researched the qualities of exhibits that visitors like in "Characteristics of Ideal Museum Exhibits." Formative evaluation became a stronger force during the 1980s, and Griggs and Manning (1983) showed that the rough mock-ups yielded reliable predictions of the final exhibit's performance. Currently, Miles and staff continue the search for educational principles and theories that are truly applicable to exhibit design. Their extensive work with a combination of methods and dedicated focus on the lay public as communication-receivers has provided examples for practitioners in the U.S.A. and other countries. See, e.g., Alt, 1979, 1980; Griggs, 1981; Miles & Alt, 1979; Miles & Tout, 1978; Miles and McManus, 1993; Jarrett, 1986 and Miles, 1986, 1988, 1993. Miles' wit and lucidity has helped keep the museum studies field humane, balanced, and on task—improving the effectiveness of educational exhibits.

The Smithsonian Institution has had a long history of institutional commitment to evaluation and impact studies, with some notable variety in terms of philosophy and methodology. In the early 1970s, Chandler Screven, from the University of Wisconsin-Milwaukee, was invited to give a series of lectures and to conduct studies of visitors using the current methods from the psychology of motivation and learning (Screven, 1974). One of his students, Robert Lakota, was hired in 1973 to conduct research for the new Office of Museum Programs for the Smithsonian Institution. As mentioned above, Lakota's work was characterized by detailed statistical data analysis. He was replaced in the late 1970s by naturalistic researcher Robert Wolf from Indiana University, who conducted more open-ended, descriptive, narrative research for the next ten years at the National Museum of Natural History, the National Museum of American History, the National Zoological Park and other Smithsonian sites (for examples, see Wolf & Tymitz, 1979, 1981a, 1981b). One of Wolf's assistants, Mary Ellen Munley, became a prominent researcher herself, based in Washington, D.C. during the 1980s, using similar social science methodologies. Currently, the Smithsonian's in-house researcher, Zahava Doering, advocates the use of quantitative analysis, and systematic, scientific evaluation (when affordable) to provide the best answers to questions about impact (Doering & Pekaric, 1993a). Numerous individual studies contracted by the Smithsonian Institution over the years are available from the Institutional Studies Office.

At the Franklin Institute Science Museum and Planetarium in Philadelphia, Minda Borun's work as the principal investigator for research on exhibit effectiveness resulted in publication of two books. *Measuring the Immeasurable*, in 1977, was a study of the impact of a whole museum visit, and, with M. Miller in 1980, *What's In A Name?*, reported on research on the effectiveness of explanatory labels. In the first book, Borun compared the responses of pre- and post-visit groups at the Franklin Institute. Measurements included learning from exhibits, attitude toward the museum and attitude toward science. There were significant increases in scores from pre-visit to post-visit groups. Borun also found that, as a result of the visit, children in grades 7–9 earned higher scores than college students. Since most of the exhibits were designed for school-age children and since this age is one of the largest segments of the population, this finding was encouraging. The attitude-toward-museum rating scale showed decreases in both positive attitude from pre- to post-visit and positive attitudes toward science (and scientists and technology). Borun speculated that these decreases in positive attitudes may have been due to confusion and disorder resulting from ongoing construction in the museum at the time. This study was one of the first to demonstrate the overall impact of a museum visit on visitor learning. Both of these popular, available publications stressed the difficulties of doing research in museums. They were often quoted, but also criticized for the linear, cognitive focus of some of the research questions and the limitations of the use of experimental design, factors acknowledged clearly by the authors.

Two other examples of institutions with high levels of commitment to visitor studies (i.e., dedication of financial and personnel resources over a number of years) and that use studies to improve exhibits are Brookfield Zoo and the Field Museum of Natural History, both in Chicago:

- In the early 1960s, Brookfield Zoo's director, then animal research coordinator, George Rabb, did one of the first studies (somewhat tongue in cheek) of visitor reactions to zoo labels, "The Unicorn Experiment" (Rabb, 1969). For the past six years, Brookfield Zoo has had a full-time researcher/evaluator on staff. In-house publications (Technical Reports) cover several formative and summative studies which document informal education impact and visitor survey data. Other articles have been published in museum-related journals (Serrell, 1981; Birney, 1988, 1991; Heinrich, et al., 1992).
- Rather than employ one full-time staff researcher, the Field Museum has made use of consultants as contractors for visitor survey research, front-end, formative and summative evaluation, staff training seminars and symposia. Numerous in-house reports have resulted from studies associated with Field's recently renovated exhibits, including the Animal Kingdom Project, "Traveling the Pacific" (partially reported on in Hayward & Rabineau, 1989) and "Life Over Time," as well as the new exhibit, "Africa." One of the most complete examples of using evaluation as an integral part of the exhibit development process was during the renovations of the American mammal diorama hall, resulting with the exhibit, "Messages from the Wilderness." A summary of that process is being published, and there are plans to continue and intensify future visitor research efforts at the Field Museum.

The Exploratorium in San Francisco has had a long tradition of evaluation that differs in several ways from the examples cited above. Exploratorium exhibit designers themselves work independently to create exhibit elements that demonstrate physics phenomena, and do "visitor studies" that are largely unstructured and not related to specific learning outcomes. Rather, the emphasis has been on watching how visitors used elements and making changes based on those observations. The main goal is to make elements that function well (e.g., don't break down) and that visitors can intuitively figure out how to use correctly (e.g., by building a foot rest on the rotating gyro chair, exhibit developers kept visitors from dragging their feet that slowed the spinning). The Exploratorium has published several "cookbooks" of exhibit construction plans for the replication of their exhibits. The cookbooks informally allude to findings from the testing of exhibits mainly as they relate to improvements in construction, maintenance, durability, and visitor safety.

Systematic data collection, analysis, or formal reports from the Exploratorium about visitor behavior or actual impact are rare in the museum literature, although there are many references to its open-ended potential for impact (e.g., Duensing, 1987).

Full-time professional evaluators are on staff at a relatively small number of museums in the U.S.A.: the Denver Museum of Natural History, Adler Planetarium (Chicago), American Museum of Natural History (NYC), Brookfield Zoo, J. Paul Getty Museum, and Franklin Institute. In-house evaluation staff have the advantages of being close at hand to respond to immediate needs and having familiarity with the processes within the "culture" of the museum. Advantages of using outside consultants are that they can bring a fresh perspective and do not have to be involved in ongoing internal politics (such as territoriality, and ego struggles). The CARE Consultants Directory lists 35 people who are skilled in and available for doing visitor studies.

Institutional commitment to evaluation does not lie with the edifice. Rather, it lies in the hands of the current administration, and management can change. When the managers are not committed to making visitor studies a part of the process, it is difficult for other members of the staff to get the resources necessary to do them. If, for the first time, management becomes concerned with rigorously gathering visitor feedback data, some staff get defensive.

Impact Studies by Academics

Many of the studies about learning impact in museums have been driven by academic interest in visitor studies. Active in the field of museum studies as teachers and researchers are Ross Loomis (Colorado State University) and Chandler Screven (University of Wisconsin-Milwaukee). Screven's work focuses mainly on the study of motivation and communication among voluntary learners and how to create impact through the application of educational psychology methods in the exhibit setting (Screven, 1974, 1986, 1990, 1991). Both Loomis and Screven are mentioned elsewhere in this chapter.

Stephen Bitgood, at the Psychology Institute at Jacksonville State University, Alabama, has involved more than a dozen students in visitor behavior research. The nearby Anniston Museum of Natural History has been a training site for students, and numerous theoretical, empirical and methodological studies were done there, many of which resulted in reports available from the Center for Social Design in Jacksonville.

At the University of California, Berkeley, Mac Laetsch encouraged students to study visitors in museums using non-experimental research techniques. Several of them used nearby sites to conduct their studies—the San Francisco Zoo (Gottfried, 1979, Rosenfeld, 1979); Lawrence Hall of Science and the Exploratorium (Diamond, 1986; Laetsch, W., et. al., 1980) and Steinhart Aquarium (Taylor, 1986). Their studies offer excellent examples of the value of descriptive research in visitor learning using the ethological approaches of observation and data gathering that have been used extensively in observing animal behavior in the wild. Diamond (1982) argues that the purpose is to study behaviors that occur naturally in an environment rather than behaviors that are "elicited experimentally by the researcher." Data collection attempts to include a large variety of behavioral categories with empirical rather than theoretical derivations. As of 1993, Judy Diamond and Sam Taylor held administrative positions in museums and were strong advocates for visitor studies.

Researchers at the University of Florida, using Florida State Museum as a research site, have pursued cognitive theories of learning in museum settings, emphasizing mental processes such as attention, memory, reasoning, and problem solving. Cognitive development, information processing, mental model and constructivist approaches and learning styles have been studied by M. L. Koran and J. J. Koran, Jr. (1979). The Korans and several of their students (some who have since become active museum practitioners and visitor researchers, e.g., Lynn Dierking, John Falk, John Scott Foster) have conducted research in museums (Koran, J. J., et al., 1983; Koran, J. J., et al., 1986; Koran, J. J., et al., 1988).

Studies based on theoretical paradigms have come from academic settings more often than from the museum practitioners. One notable exception to this is Borun's theories about naive notions in science, that is, visitors' widely shared misunderstandings of scientific principles. For example, Borun and Massey (1990) examined common misconceptions about gravity, including: gravity is air pressure, or requires air to work; gravity is generated by the rotation of the earth; gravity is due to the orbit and position of planets in our solar system; gravity is magnetism; and, gravity comes from the sun. An exhibit device was developed that allowed people to experiment with a ball falling in a tube, with and without air, to allow the visitor to see that the ball falls in either medium. This device significantly decreased two of the five common misconceptions and increased understanding of an accurate concept: Gravity is related to mass. The percentage of visitors who expressed this increased from 36 to 57%.

Another exhibit device was developed (Borun & Adams, 1991) to address the misconception that the earth's rotation generates its gravitational pull. The device required several modifications in order to significantly reduce the misconception. One form of the label that stated, "Spinning does not create gravity" was not successful and the investigators suggested that the word "not" was often ignored, and people tended to read what they wanted to substantiate their misconception. Only when visitors are brought into an active process, creating a loop "sending the visitor back to the device to test the information, then back to the label to interpret the experience" (1991, p. 118) will visitors be able to establish a dynamic relationship with the information—a prerequisite for changing a misperception. The three-year project is summarized in *Curator* (Borun, *et al.*, 1993).

More Integrated Approaches to Studying Impact

The behaviorist models of learning that shaped many of the earlier studies in museums used separate questions to tap into visitors knowledge and attitudes as visitor researchers attempted to understand how visitors learn from exhibits. The use of measures which capture a range of impacts as they occur together naturally has become more popular as other psychological perspectives have been integrated into the kinds of questions researchers ask.

- Paulette McManus has written extensively on the complex interactions of how visitors' perceptions and use of exhibits are influenced by their social group composition (e.g., adults with children, adults alone) and by exhibit design—particularly the use of text in labels (1987, 1988a, b, 1989a, b, 1990, 1991a, b). McManus argues convincingly for the integration of cognition and affect in evaluations of impact in informal learning, saying that "A bifurcated approach . . . is harmful and unrealistic, as some museum staff are beginning to protest." (1993, p. 108).
- David Uzzell (1993) reviews 15 years of summative evaluation in the Department of Psychology at the University of Surrey and describes the strengths and weaknesses of the three approaches: behaviorist studies, cognitive studies, and socio-cognitive studies. He concludes that equal attention should be given to the assessment of exhibit media's effectiveness in creating and enhancing learning and to the visitors' own active constructions and interpretations in the social and educational contexts that are present.
- Bernard Schiele's work in progress, "Creative Interaction of Visitor and Exhibition" (1992), examines the transformations of our perceptions of visitors, the role of communication in exhibitions, and the

relationships and interactions of evaluation. His analysis rises above the familiar arguments about schools of thought and methodology and takes a historic and developmental look at the concept of visitors and "raises an inquiry on one of the integrating poles of contemporary museum discourse."

Impact Studies in Art Museums

The study of informal learning in museums is informed by many sources of helpful information—theories, studies, opinion papers—that lie out-side the realm of museum literature. In fact, much of the information comes from other sources, such as educational technology, communications theory, and mass marketing techniques, to name a few. To learn about how people learn in science museums without first looking at other types of museums, however, such as history and art, would be to ignore an obvious resource. Impact studies conducted in art museums that have direct implications for science museums primarily involve the way both art and science conjure up feelings of *intimidation* in visitors to those museums. The following publications discuss this and other issues:

- "Staying Away: Why people choose not to visit museums" by M. G. Hood (1983) explored the characteristics that people want and value most in their leisure activities. Some of the attributes that non-museum-goers desire, that might be missing in many museum settings, include family oriented, social experiences in a setting where they feel competent, in control and not overwhelmed. Although Hood's study dealt with art museums, there are similar situations in many other exhibit settings.
- *The Denver Art Museum Interpretive Project,* by Melora McDermott-Lewis (1990) summarizes a two-and-a-half-year effort to develop a conceptual framework for creating interpretive materials for novice, or lay visitors. The results of 13 experimental label and gallery guide projects are described. The interpretive experiments involved cru-cial visitor feedback using a variety of evaluation strategies. Their findings through "small-sample techniques" will be of interest to museums with similar time and resource limitations. Ross Loomis, from Colorado State University, who has had a long-term involve-ment with museum studies and museum staff development, helped with the interpretive project, as well as many other studies at the Denver Art Museum. (See also Loomis, 1988.)

- *Insights*, published by the Getty Center for Education (1991), is a comprehensive study of visitors' attitudes and expectations. It contains summaries and discussions about focus group studies conducted at eleven different art museums. What these institutions heard visitors say and how they are responding to the issues and concerns voiced by their visitors will be of interest particularly to natural history museums or collection-based science museums. Insights from people with interesting perspectives from outside the museum field included are those of Neil Harris and Milhaly Csikszentmihalyi. At a price of $5 per copy, it is without a doubt the most affordable visitor studies literature resource at this time.

Other books useful to the study of learning in museums that are visitor oriented and emphasize the achievement of positive impact include *Museum Visitor Evaluation*, by Ross Loomis (1987); *The Design of Educational Exhibits*, compiled by Roger Miles *et al.* (1988); and *Planning for People in Museum Exhibitions*, by Kathleen McLean (1993).

New works are being generated every month. To help the reader keep up with these visitor studies publications, good bibliographies are available. In addition to *Visitor Studies Bibliography and Abstracts* (1993), edited by Chandler Screven, is *Recent and Recommended* (1991), edited by Kathleen McLean, published by National Association for Museum Exhibition. Anyone interested in impact should try to keep abreast of the literature and become their own critic of it.

> *One of the things I have been struck with . . . is how inarticulate we all are, because we do not have very good visual or verbal or other models to describe the transactions and interactions of informal learning.*
> —M. Templeton, as quoted in St. John & Perry, 1993

Issues that Have Slowed Change

After a surge of interest in visitor research and impact studies in the mid-1970s, certain issues tended to slow progress and inhibit the institutionalization of methods and procedures.

In the history of visitor studies, linguistic and conceptual issues continue to be discussed, sorted out, refined, and changed. No theories or practical advice for the field have sprung into consciousness fully formed, clearly defined, and universally accepted. Rather, the process is one of muddling: slowly dealing with the issues when time and funding allow, making progress, falling short of lofty objectives, arguing, and making mistakes.

Many misunderstandings between proponents of different schools of thought can be traced to ill-defined vocabulary, frequently involving the terms "research" and "evaluation." They are not always interchangeable notions. Although the data collection strategies are often the same, research and evaluation have often different goals, purposes, and different questions. *Research* seeks to generalize, to find cause and effect, and investigates questions about "How does it work?" and "Why does it happen?" under more controlled conditions than *evaluation*, which tends to be specifically goal-referenced to a problem at hand, "Did it work?" (For more discussion about the differences and misunderstandings, see Miles, 1993; and Shettel, 1990.) The confusions in meanings have led to unfair criticisms of evaluation studies based on the standards of research practices.

Another source of confusion is the definition and distinction between qualitative and quantitative approaches. The confusion can be minimized by referring not to qualitative or quantitative evaluation or evaluators, but to qualitative and quantitative methods of data analysis. A review of the literature suggests that the majority of visitor professionals use both qualitative and quantitative methods in visitor studies depending on the nature of the questions being asked. It's not an either-or situation.

Not being familiar with the literature has caused problems. Professionals from inside and outside the field (e.g., from formal education, art history, or marketing) find it easy to give opinions about what is wrong with educational effectiveness in museums and how it might be "fixed." Naive notions within the profession and a historical approaches from people who have not done their homework have led to many articles and sessions at conferences where authors/speakers are reinventing ideas and proposing solutions they think are original, when they could be building on existing ideas and helping advance them further. There are no excuses for not being familiar with the literature, not with the annotated, indexed, up-to-date *Visitor Studies Bibliography and Abstracts* available.

Using the wrong methodology for the wrong situation has been the source of many criticisms of evaluation and impact studies in museums. In particular, formal education research strategies are bound to be difficult and inappropriate in settings that lack structure and contain a diverse audience, as is often the case in museums.

- One of the most problematic issues of methodology is that of using written verbal formats to investigate impact with museum audiences (e.g., questionnaires, interviews, tests). Informal learning doesn't always promote articulate, immediate verbal formulations. Visitors often find it hard to say what they expect, experience, or learned from a brief encounter with an exhibit element. More experiments with different techniques to help visitors articulate their mental connections are needed (e.g., taking pictures, drawing pictures, using a tape recorder for thinking out loud).

• Grants from foundations that support informal science education programs and require evaluation have been both a friend and a foe of visitor studies. While insisting on educational evaluation as part of the grantee's program can increase the likelihood of positive impact, the perpetuation (real or imagined) of an inappropriate or limited methodology of experimental design formats (e.g., pre- and post-tests, control groups, cognitive behavioral objective emphasis) while looking for impact has stifled other new creative solutions for the study and documentation of visitor behavior.

Another muddle is over how "informal" museum learning is. The museum tends to be formal, but the learner isn't. Museums have high cultural status in society, an educational mission, and organized and deliberate methods of cultural expression through exhibits. Special routines of information transmission, such as identification and interpretive text labels, cases, models, dramatic lighting, etc., are the "language" of exhibitions. Visitors, on the other hand, have all the characteristics of informality inherent in leisure activities in museums carried out with few rules, little time, lots of competition for attention, a personal and usually social agenda and variable "museum literacy" to decode the exhibit language. As museums strive to attract and serve broader audiences, people who have even less practice with using exhibits will be visiting. Orientation, self-guiding aids, and effective exhibit communication will become even more critical and important, not less.

The false dichotomy of values between educational effectiveness vs. entertainment has bogged down discussions. Both should be acting together, because most visitors will not be attracted to exhibits that appear text-heavy, pedantic, overwhelming, and neither fun nor interesting. Visitors are not knowledge seekers and didactic information transfer not a good model for making visitor-centered exhibit experiences. Visitors cannot be *forced* to learn anything. On the other hand, to accept the assumption that most visitors are incapable or uninterested in learning about the principles of science sabotages striving for greater impact and allows the justification of the existence of any exhibit, no matter how ineffective it is.

Criticism of the so-called behaviorist methods, denial of the usefulness of behaviorism theories, and dismissal of all behaviorist-like visitor studies has slowed progress. Stereotyping of the behaviorists has caused confusion for museum staff who have wanted to start to do visitor studies but did not know which method to choose—as if there were a black and white choice between treating visitors in a coercive, calculated scientific method, like they were rats in a maze, or treating them "naturally," responding sensitively to their individual personalities: another false dichotomy.

All theories are flawed, but that does not mean that they are not useful. By being hesitant to use imperfect paradigms or becoming paralyzed with a fear of error and bias, museum practitioners can do the worst: nothing. As Nigel Norris (1993) recommends: know your biases, don't be ignorant of them, proceed to do your best, share what you know, and allow for criticism.

The introduction of new paradigms and new metaphors alone won't change the problem. Instead, they may help divert attention and maintain the worst part of the status quo—designing, building and traveling exhibits that don't communicate as well as they could.

In looking for impact, some people want more anecdotes about visitors behavior; others long for long-term evidence. A case can also be made for the need to look more closely at the seemingly mundane: immediate observations of visitor behavior. We haven't yet thoroughly explored short-term impacts and become articulate about them. What do we really know about learning that takes place in seconds, not semesters? There is a need for more naturalistic dialogue with visitors to hear about their expectations and instantaneous meaning-making in exhibitions to provide the necessary groundwork of information for theory-building about the conditions and nature of impact. We need more thorough descriptions and demonstrations of our assumptions, such as the existence of "the more serious, interested visitors," and the "bimodal response," what makes "successful exhibits," and "empowered visitors," or the importance of "infrastructures."

Part of the problem is often a lack of agreement at the outset of exhibit planning on what the expectations for impact might actually be. Thus, the focus is on the intent, the offering, but not the results. Instigation, funding, and praise for exhibit ideas are usually on the basis of "a neat idea" in and of itself, ignoring the importance of evaluation feedback and impact studies. Did the neat idea work? If not, why? Failure is an important lesson that is not often shared.

The lack of focus on impact at the outset often means that standards for impact are very low. For example, "If but one visitor leaves an exhibition with a new sense of wonder, understanding or useful purpose, that exhibition can be said to have succeeded," claims M. Belcher in *Exhibitions in Museums* (1991). Then, there are those who believe it is inappropriate to have any standards, who would rather have visitors be confused than set any criteria for measuring specific learning goals, or those who see goal-related measurement as a form of coercion. The AAM Curators Exhibit Award has no criteria for judging impact, although they mention but do not define success with the audience. As long as "successful" is a subjective opinion, exhibit impact will remain a popularity contest. Shettel (1973) addressed this issue in his now classic paper in *Museum News*, "Exhibits: Art form or educational medium?", and he offered many strategies for determining exhibit effectiveness (Shettel *et al.*, 1968) such as attracting, holding and teaching power. While those terms have become part of the vocabulary to describe impact, they have not been widely adopted as standards.

Criticism is warranted of exhibit impact studies that have gathered data unsystematically, used non-random samples, based statistical conclusions on small sample sizes, and that over-generalize their findings. (For example: 15 people—8 staff members and 7 visitors—are shown two samples of label text design and asked which one they prefer. Ten visitors say label A, and the evaluator concludes that design A will work best for all labels.)

Criticism for being a useless study is warranted—from the museum practitioner point of view—of studies that may have followed the right methodologies but asked an uninteresting question, conducted the study under unrealistic conditions with a limited segment of the audience, and used only one type of measuring strategy: (for example: the effect of recessing a beetle display on 7th and 8th grade students, measured with a 25-item multiple-choice test.)

Will there ever be agreement in the field of visitor studies on the right thing to measure and the right way to measure it? Probably not, but there really doesn't need to be for the field to make progress.

The museum field is fraught with an interplay of factors deriving from contradictory discourses. . . .

—B. Schiele, 1992

Signs of New Synthesis

Despite criticisms and confusing issues, there are many signs of new synthesis, clearer goals, and holistic models.
Evidence that the field of visitor studies is learning from and sharing with other fields, especially psychology and education, is in the number of citations in the *Visitor Studies Bibliography* from publications such as *Journal of Educational Psychology, British Journal of Psychology, American Psychologist, Journal of Educational Research,* and *Science Education.*

Other fields contribute as well, through publications such as *Leisure Sciences and Journal of the Academy of Marketing Science.* Closely allied but sometimes overlooked in the past are the groups associated with outdoor recreation and environmental education, including the National Park Service, *The Interpreter, Proceedings of the Association of Interpretive Naturalists,* and in Great Britain, the Society for the Interpretation of Britain's Heritage, and Centre for Environmental Interpretation. There has been more cross-pollination of ideas between these groups, notably in the meeting of a third global congress of Heritage Interpretation International in 1991 in Hawaii.

Management teams from museums have taken advantage of the training seminar "Quality Service" from Walt Disney Company in Florida. Training for good client management is consistent with the emphasis and importance museums are placing on visitors' needs. Help also has come from consultants with film and entertainment expertise, and influences on museum architecture have been made by companies that develop marketplaces (e.g., waterfront shopping places, outlet malls).

As the museum practitioners themselves gain more experience, they propose new ways of looking at the complex inter-relationships of exhibitions and visitors. John Falk & Lynn Dierking's popular book, *The Museum Experience* (1992), suggests a framework of three contexts—personal, social and physical—to describe and understand the visitor's perspective of a visit to a museum. The authors of this chapter are working independently on other, different configurations: Bitgood is organizing a framework of visitor variables, exhibit variables and interactions between the two as "An Integrative Approach to Visitor Learning"; Serrell is collecting and analyzing time data from more than 25 separate studies of visitors' use of exhibits to create a model of time patterns across types of exhibits (varying in subjects, sizes and number of elements) in "The Momentary Shrine: A New Look at Visitors, Exhibits and Time."

A recent visitor studies conference was organized by the Science Museum in London, a stimulating assembly of practitioners and academics. This meeting resulted in the publication *Museum Visitor Studies in the 90s* (S. Bicknell and G. Farmelo, Eds., 1993), which contains some frank and lively discussions about many of the issues and ideas contained in this chapter.

As museum decision-makers (administrators) are seeking information that will help make museums more educationally effective, and practitioners (model-makers) are providing better definitions of the processes, products and impacts of exhibits, the field will evolve. The trend is toward multi-method approaches to studying visitors, triangulation of data analyses, and meta-studies involving multiple institutions and study sites. One such study (although it fell short of its original, ambitious goal of developing a whole new concept, theory, and methodology of evaluating exhibits) is "The Museum Impact and Evaluation Study" (MIES) coordinated with nine museums by the Chicago Museum of Science and Industry (Anderson & Roe, 1992).

The emphasis on the importance of studying and valuing visitor characteristics, such as learning styles, levels of interest and involvement will continue. In turn, the characteristics of a successful exhibit, that is, one that is popular, engaging, social, participatory, allowing visitors to express and exercise their own uniqueness and personal agendas, will also develop. With that knowledge, standards for success in terms of use and outcomes can be derived, tested and refined.

More experiments with different techniques to help visitors articulate their mental connections are being tried. Maybe visitors can't tell exactly what they learned, but they can tell something about what they did, whether they enjoyed it and found something meaningful, and through triangulation with other methods, such as tracking and timing, self-reported data can be cross-checked with what visitors did with their feet.

Trying to capture visitors' reactions in less verbal, more open-ended techniques to get at what's going on inside their heads has always been a goal of visitor studies. The use of less structured, less closed-question methods, such as unobtrusive video taping, self-taping with a hand-held recorder, self-documenting with a Polaroid camera, drawings and content analysis of dialogue are being used more frequently. Some practitioners are less inhibited about experimenting with less than perfect methodology than others. Still, certain standards should be followed to be acceptable, and the Committee on Audience Research and Evaluation of AAM outlines some of these responsibilities and competencies in their

Professional Standards (see *ILVS Review*, 2(2), 1992). These guidelines guide, but do not constrain.

The ideal scenario that is emerging for informal science education in museums goes something like this: The exhibit environment promotes, through exhibit variables (e.g., objects, media, layout), the highest possible level of successful intended outcomes (i.e., exhibit developers' specific communication objectives) and minimizes the incorrect or unintended visitor outcomes (e.g., reinforcement of a naive notion, finger pinched in an interactive device) while at the same time, it allows visitors opportunities to explore and enhance their own characteristics (e.g., interest, prior knowledge, curiosity) and personal objectives (e.g., social, recreation, enlightening entertainment), with the "dialogue" between the visitor and the exhibition being dynamic, not one-sided.

In the more holistic conception, the goal is to have a balance between what information and experiences visitors will find personally relevant and how museums can make that information and those experiences easily accessible while achieving their own objectives. Striking this balance focuses attention more on visitors than on content. Isn't it about time?

> *Scientists are not missionaries, and visitors are not natives to be clothed with righteous learning.*
>
> —W. M. Laetsch, 1991

What's Needed for the Future

What the future holds and needs: making museums less formal, more accessible, redefining "learning" and doing more research and evaluation.

There are two strong factors that are motivating museums to make their exhibits more physically accessible and popularly appealing: the law and the economy.

- Americans with Disabilities Act (ADA) says that museum exhibits must be accessible to visitors with hearing, motor and visual impairments. Changes in exhibits to meet ADA requirements for an impaired minority are highly likely to benefit other adult visitors (TABs, the "temporarily-able-bodied") as well as others with height-related limitations, and children.
- Museum management is looking for new ways to market museums to new audiences. Whether it is altruistic or simply economic, museums will continue to be more multicultural. However, getting

new, "underserved" visitors in the door is only the first step; the visit must be a positive and intellectually accessible experience for them to come back.

Museums that currently do not have an interest in visitor impact studies will become inspired to be more familiar with them as they grapple with these factors.

As museums continue to find new ways to be more physically and mentally accessible, they should also shift the emphasis to asking a better, more realistic question for studying impact. The question "What CAN visitors learn from this exhibit?" is a better question than "What DID they learn?" because visitors are not under any formal obligation to do anything. When they can learn, how do they do it? What things have a positive effect on learning? Models that are inclusive, that synthesize information from multiple fields and examine broader definitions of learning will help to answer those questions. Museums should make the question "What can they learn?" an empiric, evaluation query to be seriously and systematically investigated in all museums, not just a question to muse over, guess at, or hypothesize about in the abstract.

Amassing larger data bases from a variety of different museums, visitor research will allow cumulative studies to help in the search for greater patterns, trends and truths. Computer-assisted data organization and cross referencing of data bases will be in a more streamlined form with increasingly available powerful hardware and software. By having more explicit, shared strategies for carrying out visitor studies, this data combining will be possible. The basic rules of random sampling, systematic collection and quantitative analysis (but statistical treatments only when sample sizes are large enough to warrant it) will help reduce error and bias. Use of naturalistic methods will help gain insights into qualitative aspects of impact and provide a more holistic view of visitor behavior. Multiple-site, shared-strategy visitor studies can be defined by subject (e.g., natural history dioramas), patterns of use (e.g., time spent), visitor type (e.g., families), or other sub-units of interest.

Given the many time and financial constraints in museums, evaluation studies often have to be small scale. Small sample sizes (n = 25–50) can still provide useful data for the museum, but they may not be convincing or appropriate to publish as research impact studies. If the study is focused on a more theoretical question about the behavior of the users (a research-oriented study), not the efficacy of the exhibit (an evaluation study), then larger sample sizes (n = 100+) are desirable and appropriate. This is also true for visitor surveys of demographics and psychographics. The intent of each study should be made clear and then it should be judged on those merits.

The sub-grouping or splitting of data—by age, gender, race, group type, prior knowledge, typologies, or styles—is often not statistically valid for small studies. Splitting becomes less important when the general evaluative goal is to reach more diverse audiences and to be successful with all types. Sub-groups are important to think about when *planning* exhibits, to build-in ways to meet a variety of needs and preferences, but can be less important as a priority in evaluation.

Many impact studies have tended to be concerned with evaluation and the improvement or enhancement of impact through making changes to exhibit elements (e.g., improved labeling, improving a brochure guide) during exhibit development. Some museums are starting to set aside ten percent of the exhibit budget for improvements after opening. Whether this is realistic remains to be seen, as many exhibits go over budget and eat into the remedial funds before revisions can be made. It is still a good idea to do some studies after opening with the intention of using the information to fix what needs fixing.

More studies are needed of both long-term impact and theoretical issues. The amount of time and money devoted to those questions by museums so far has been very small, probably because the lack of quick feedback and immediately useful data makes them less easy to justify.

Instead of isolated, a historical approaches to visitor studies, the next decade will be characterized by more "generous, broad, lateral thinking" (Bicknell, 1993) and convergence. Knowing the strengths and weaknesses of the researcher, having peer and public review of data and findings, and getting more feedback from the users of the study's findings can make studies more worthwhile.

More funding is essential to provide the resources for doing visitor studies of all sizes. Basic descriptive, ethnographic research is needed to understand the "natural history" of the museum visitor. More funding is also needed for the dissemination of what we know now, so that useful findings are more accepted and integrated. For example, despite ninety years of saying that labels should be short enough (few words) and big enough (large print) to read, and relevant (see Chambers, 1993), there are still too many examples of exhibit texts that most visitors either can't see easily, don't want to take the time to read, or can't understand.

Impact studies have often been muddied because they have looked for impacts on learning from ineffective exhibits. Before results can reveal why something works well when it does, comparative or experimental conditions need to be created and tested with whole exhibitions that are recognized as exemplary, not ones with obvious problems. For example, a study at the National Museum of Natural History on a coral reef exhibit was done by Falk (1993). The exhibit had been evaluated earlier by Wolf (1981b), who described its successes and deficiencies. During Falk's study, visitors' traffic flow and attention to exhibit elements were compared under a structured and unstructured layout. The display had a very broad topic ("What is an Ecosystem?"), and most of the 12 elements didn't attract half of the audience under either condition. It would have been more valuable if the museum chose to study an exhibit such as "Darkened Waters: Profile of an Oil Spill" that had a far more successful rate of capturing and holding visitors' attention and communicating its various messages (Pratt Museum, 1992). It would be interesting and useful to compare visitor reactions to "Darkened Waters" as it is traveled. Do visitors respond in the same way at a children's museum in Iowa as they do in a science museum in California, or a natural history museum in Virginia? The American Museum of Natural History is in the process of doing just such a study with their "Global Warming" exhibit. The National Zoological Park has done a multi-site study of their Reptile Discovery Centers at four different zoos, using pre- and post-tests, demonstrating positive impacts on the amount of time spent, learning about and more positive attitudes about reptiles for visitors at all four sites (Doering, *et al.,* 1993b). More comparative studies like these are needed.

More projects that create good models of exhibits and assessment are needed. Exhibits that increase visitors' access (physical and conceptual), demonstrate positive impacts through evaluation, and disseminate their findings through publications and traveling exhibits will help others follow suit. Boston Museum of Science's *New Dimensions for Traditional Dioramas* and National Zoological Park's *Frogs, Families and Fun,* described earlier, are two good examples of shared model programs. An innovative exhibit technique was shared when Bud Wentz created the "Easy-View Wentzscope" for the New York Hall of Science's microbiology exhibit—an exhibit element that gave people access to the microscopic world in museums everywhere.

Museum practitioners, educators, evaluators, and critics of visitor learning impact studies need to have more faith in the ability of visitors to learn, be more critical of the status quo, and to have more optimism about new possibilities as we continue to muddle our way along, tweaking methods, dealing with messy data, asking hard questions and having fun.

To empower the visitor with greater understanding is one of the most worthy aims a museum can have. For science museums to achieve this, there needs to be a conscious move forward in the content and quality of provision. Some creativity, humanity, humour, honesty, discussion, and a little humility would not go amiss.

—Gaynor Kavanagh, 1992

Summary and Final Thoughts

In summary, what do we know about the impacts of informal learning in science museums, zoos and aquariums, and where should we go from here?

1. That there are some impacts, and they are intellectual, emotional and physical, planned and unplanned.
2. That orientation, both psychological and spatial, is a very important factor that can influence impacts, positively and negatively.
3. That impacts are socially influenced and enhanced most positively by the social characteristics that are appropriate to informal learning such as interaction, sharing, parental guidance and intimacy between visiting group members.
4. That impacts are environmentally influenced and enhanced, most positively by exhibit characteristics that are suitable to informal learning settings, such as concrete and experiential activities, reinforcement of concepts and efficient communication techniques.
5. That measurements of specific impacts with the traditional tools of experimental design are often inappropriate for the confounding variability of informal settings, making the results of such assessment often disappointing or insignificant.
6. That impacts can be positively enhanced by using visitor feedback during the planning and development stages of exhibit design through front-end and formative evaluation.
7. That evaluation is essential to increasing the success of informal science learning in museums.
8. That future research on impact in museums needs to combine multiple, systematic methods and strategies that are appropriate to the voluntary, social, intrinsically motivated experiences that visitors have.
9. That there is a lot of room for improvement, even though visitors are coming to museums in droves and rarely complain.

10. That improvement in the amount of impact on informal science learning in museums—and its objective appraisal—is essential if museums are to be held accountable to their claims of having an educational role in society.

Final Musings

Why is it that when small gains in learning were first reported in experimental exhibit research, and authors like Weiss and Boutourline (1963) clearly laid out many of the shortcomings of the ability of exhibits to communicate knowledge, the whole museum community didn't react with surprise, chagrin, and corrective action? Instead, some responses were defensive and the bearers of the bad news were attacked (Wrong methodology! Coercive standards!).

How many of the visitors who are having so much fun in science playgrounds are really grasping the basic scientific principles of vision, sound, energy, and gravity as museums claim in the ads? Could they if they tried? If museums were more willing to admit that they don't often live up to the marketing hype, would visitors stop coming to museums? Probably not. In fact, actively involving visitors in the process of renovation, rejuvenation and improvement of exhibits through evaluation can help visitors find a new sense of meaning and personal ownership for more of the museum's exhibits, rather than having just a respectful reverence for something they really don't understand.

If museums are to be accountable for providing what they say they are providing, they must measure *something*. "Visitors will learn. . . ," "Visitors will discover. . . ," "Visitors will understand. . . ," are still very common ways of describing what will happen to museum-goers. Redefining impact, such as "visitors will be offered rich and inviting high quality experiences," (St. John & Perry, 1993) does not change the problem, it just raises the rhetoric to a new plateau. Museums are still faced with measuring impact: What is "rich"? How "invited" did visitors feel in the exhibit? If museums take money to help fund an educational endeavor, there is an extra responsibility attached to show what happened, no matter how it is defined.

By using research and evaluation techniques that are adaptable and appropriate to the museum's informal setting and self-motivated learners, the results of impact studies are bound to be better, especially if the exhibit was developed with appropriate objectives, appropriate language and evaluated with visitor feedback during the exhibit development process.

It is not enough to read, learn from, and apply the findings of other peoples' studies. Each new exhibit situation is a new combination of many variables which necessitates the use of formative evaluation to make it work. For example, studies have shown that good orientation can prepare visitors and reinforce their understanding of an exhibit topic and improve learning. Designing and installing an orientation sign for the exhibit is not enough. Orientation labels are typically one of the elements most ignored by visitors! To be effective, orientation has to be *used*. To make sure that the orientation sign is the right size, in the correct location, attractive to visitors, legible and comprehensible, it probably needs to be mocked up and tested before it is finally built and installed.

There is a lot of room for improvement. The best place to start making a realistic attack on the problem (failure of an exhibit element to communicate) is at the point where a self-selecting visitor chooses to stop and look and read and do. At that moment, is the visitor's experience likely to be one that is personally relevant, interesting, accomplished with ease, and encouraging of social interaction? Will they feel magic, passion, and awe? If it is a structured hands-on activity, does it involve looking closely, making choices, doing something in a logical step by step process, with guided discovery, and a chance to test yourself and feel competent? If it is an unstructured "phenomenarium," is the cognitive complexity manageable? Is guidance available for those who seek it? Are alternate learning styles provided for? If all these things are happening, there's probably impact. And it's measurable. Museums have the reputation and responsibility to do a better job. Muddle forth!

References

Abler, T. S. (1968). Traffic patterns and exhibit design: A study of learning in the museum. In S. De Borheghyi & I. Hanson (Eds.), *The Museum visitor, Publication in museology: No. 3.* Milwaukee, WI: Milwaukee Public Museum, 103–141.

Alt, M. (1979). Improving audio-visual presentations. *Curator*, 22(2), 85–95.

Alt, M. (1980). Four years of visitor surveys at the British Museum (Natural History). *Museums Journal*, 80, 10–19.

Alt, M., & Shaw, K. (1984). Characteristics of ideal museum exhibits. *British Journal of Psychology*, 75, 25–36.

American Association of Museums (1992). *Excellence and equity: Education and the public dimension of museums.* Washington, DC: American Association of Museums.

Anderson, P. & Roe, B. (1992). *The museum impact and evaluation study.* Chicago Museum of Science and Industry.

Anderson, R. C., Pichert, J. W., & Shirey, L. L. (1983). Effects of the reader's schema at different points in time. *Journal of Educational Psychology*, 75(2), 271–279.

Association of Science-Technology Centers (1993). *What research says about learning in science museums, vol. 2.* Washington, DC: Association of Science-Technology Centers.

Barnard, W. A., Loomis, R. J. & Cross, H. A. (1980). Assessment of visual recall and recognition learning in a museum environment. *Bulletin of the Psychonomic Society,* 16(4), 311–313.

Belcher, M. (1991). *Exhibitions in museums.* Leicester & London, England: Leicester University Press. [Also, Washington, DC: Smithsonian Institution press.]

Bicknell, S. (1993). *"Goal-referenced, goal-free or own goal?"* Paper presented at Museum visitor studies in the 90s conference, London.

Bicknell, S., & Farmelo, G., Eds. (1993). *Museum visitor studies in the 90s.* London: Science Museum.

Birney, B. (1988). Brookfield Zoo's Flying Walk Exhibit. *Environment and Behavior,* 20(4), 416–434.

Birney, B. (1991). An evaluation of bird discovery point's docent demonstration area. *Current Trends in Audience Research & Evaluation.* Washington, DC: AAM Committee on Visitor Research and Evaluation, 9–13.

Bitgood, S. (1990a). *The role of simulated immersion in exhibition.* Technical Report No. 90-20. Jacksonville, AL: Center for Social Design.

Bitgood, S. (1990b). The ABCs of label design. In S. Bitgood, A. Benefield, & D. Patterson, (Eds.), *Visitor studies: Theory, research and practice, volume 3.* Jacksonville, AL: Center for Social Design, 115–129.

Bitgood, S. (1991a). Suggested guidelines for designing interactive exhibits. *Visitor Behavior,* 6(4), 4–11.

Bitgood, S. (1991b). What do we know about school field trips? *ASTC Newsletter,* Jan/Feb, 5–6,8. No. 14, What Research Says. . . .

Bitgood, S., Benefield, A., & Patterson, D. (1989). The importance of label placement: A neglected factor in exhibit design. In *Current trends in audience research.* Washington, DC: AAM Visitor Research and Evaluation Committee, 49–52.

Bitgood, S., Conroy, P., Pierce, M., Patterson, D., & Boyd, J. (1989). Evaluation of "Attack & Defense" at the Anniston Museum of Natural History. In *Current trends in audience research.* Washington, DC: AAM Visitor Research and Evaluation Committee, 1–4.

Bitgood, S., & Loomis, R. (1993). Introduction: Environmental design and evaluation in museums. *Environment and Behavior,* 25(6), 683–697.

Blackmon, C. P., LaMaster, T. K., Roberts, L. C., & Serrell, B. (1988). *Open conversations: Strategies for professional development in museums.* Chicago, IL: Education Dept., Field Museum of Natural History.

Bloom, J., Powell, E., Hicks, E., & Munley, M. (1984). *Museums for a new century: A report of the Commission on Museums for a New Century.* Washington, DC: American Association of Museums.

Bloomberg, M. (1929). *An experiment in museum education.* (New Series No. 8). Washington, DC: American Association of Museums.

Boram, R. (1991). What are school-age children learning from hands-on science center exhibits? In A. Benefield, S. Bitgood, & H. Shettel (Eds.), *Visitor studies: Theory, research, and practice, volume 4.* Jacksonville, AL: Center for Social Design, 121–130.

Borun, M. (1977). *Measuring the immeasurable: A pilot study of museum effectiveness.* Washington, DC: Association of Science Technology Centers.

Borun, M., & Adams, K. (1991). From hands-on to minds-on: Labeling interactive exhibits. In A. Benefield, S. Bitgood, & H. Shettel (Eds.), *Visitor studies: Theory, research, and practice, volume 4.* Jacksonville, AL: Center for Social Design, 115–120.

Borun, M., & Massey, C. (1990). Cognitive science research and science museum exhibits.

In S. Bitgood, A. Benefield, & D. Patterson (Eds.), *Visitor studies: Theory, research, and practice, volume 3.* Jacksonville, AL: Center for Social Design, 231–236.

Borun, M., Massey, C., & Lutter, T. (1993). Naive knowledge and the design of science museum exhibits. *Curator,* 36(3), 201–219.

Borun, M., & Miller, M. (1980). *What's in a name?* Philadelphia, PA: Franklin Institute and Science Museum.

Brooks, J. A. M., & Vernon, P. E. (1956). A study of children's interests and comprehension at a science museum. *British Journal of Psychology,* 47, 175–182.

Carlisle, R. (1985). What do children do at a science center? *Curator,* 28(1), 27–33.

Chambers, M. (1993). After legibility, what? *Curator,* 36(3), 166–169.

Cohen, M. S., Winkel, G. H., Olsen, R., & Wheeler, F. (1977). Orientation in a museum: An experimental visitor study. *Curator,* 20(2), 85–97.

Committee on Audience Research & Evaluation (CARE), (1992). The professional standards document. *ILVS Review,* 2(2), 266–272.

Cone, C. A., & Kendall, K. (1978). Space, time and family interactions: Visitor behavior at the Science Museum of Minnesota. *Curator,* 21, 245–258.

Current Trends in Audience Research & Evaluation, Committee on Audience Research & Evaluation (CARE). Washington, DC. 1987–.

Davidson, B. (1991). *New dimensions for traditional dioramas: Multisensory additions for access, interest and learning.* Boston, MA: Museum of Science.

De Borhegyi, S. F. (1963). Visual communication in the science museum. *Curator,* 6(1), 45–57.

DeWaard, R. J., Jagmin, N., Maistro, S. A., & McNamara, P. A. (1974). Effects of using programmed cards on learning in a museum environment. *Journal of Educational Research,* 67(10), 457–460.

Diamond, J. (1982). Ethology in museums: Understanding the learning process. *Journal of Museum Education: Roundtable Reports,* 7(4), 13–15.

Diamond, J. (1986). The behavior of family groups in science museums. *Curator,* 29(2), 139–154.

Diamond, J., Smith, A., & Bond, A. (1988). California Academy of Sciences discovery room. *Curator,* 31(3), 157–166.

Dierking, L. (1989). The family museum experience: Implications from research. *Journal of Museum Education,* 14(2), 9–11.

Discover (1993) Ten great science museums. 14(11) November, 77–113.

Doering, Z. & Pekaric, A. (1993a). The exhibition dialogue: an outline. *Exhibitionist,* 12(2), 8–11.

Doering, Z., Marcellini, D. & White, J. (1993b). *Integrating informal science education into traditional zoo exhibits: lessons from the Reptile Discovery Centers.* A talk at Visitor Studies Association Conference, Albuquerque, 1993.

Duensing, S. (1987). Science centres and exploratories: A look at active participation. In D. Evered & M. O'Conner (Eds), *Communicating science to the public.* New York: John Wiley & Sons, 131–142.

Eason, L., & Linn, M. (1976). Evaluation of the effectiveness of participatory exhibits. *Curator,* 19(1), 45–62.

Edington, G., Siska, J., & Perry, D. (1993). "The intrigue is with the detail": Probing for affective responses. *Current trends in audience research.* Washington, DC: AAM Committee on Audience Research and Evaluation, 12–16.

Falk, J. (1983). Time and behavior as predictors of learning. *Science Education,* 67, 267–276.

Falk, J. (1993). Assessing the impact of exhibit arrangement on visitor behavior and learning. *Curator, 36*(2), 133–46.

Falk, J., & Balling, J. (1982). The field trip milieu: Learning and behavior as a function of contextual events. *Journal of Educational Research, 76*(1), 22–28.

Falk, J., & Dierking, L. (1992). *The museum experience.* Washington, DC: Whalesback Books.

Falk, J., Koran, J., Dierking, L., & Dreblow, L. (1985). Predicting visitor behavior. *Curator, 28*(4), 249–257.

Falk, J., Phillips, K., & Boxer, J. (1992). Invisible Forces Exhibition: Using evaluation to improve an exhibition. In D. Thompson, A. Benefield, S. Bitgood, H. Shettel, & R. Williams (Eds.), *Visitor studies: Theory, research, and practice, volume 5.* Jacksonville, AL: Center for Social Design, 191–205.

Feher, E., & Rice, K. (1985). Development of scientific concepts through the use of interactive exhibits in museums. *Curator, 28*(1), 35–46.

Flagg, B. (1991). Visitors in front of the small screen. *ASTC Newsletter,* Nov/Dec, 9–10.

Friedman, A. (1991). Mix and match. *Museum News,* July/August, 38–42.

Gennaro, E. D., Stoneberg, S. A., & Tanck, S. (1982). Chance or the prepared mind? *Journal of Museum Education: Roundtable Reports, 7*(4), 16–18.

Getty Center for Education (1991). *Insights: A focus group experiment.* Los Angeles, CA: The J. Paul Getty Trust.

Goldberg, N. (1933). Experiments in museum teaching. *Museum News, 10*(15), 6–8.

Gottfried, J. (1979). *A naturalistic study of children's behavior in a free-choice learning environment.* Ph. D. dissertation, University of California-Berkeley.

Gottesdiener, H. & Boyer, J. (1992). *Self-testing on Raphaël: How a computer stimulates visitors in an art exhibition. ILVS Review: A Journal of Visitor Behavior, 2*(2), 165–180.

Greenglass, D. I. (1986). Learning from objects in a museum. *Curator, 29*(1), 53–66.

Griggs, S. (1981). Formative evaluation of exhibits at the British Museum. *Curator, 24*(3), 189–202.

Griggs, S., & Manning, J. (1983). The predictive validity of formative evaluation of exhibits. *Museum Studies Journal, 1*(2), 31–41.

Hayward, D. G., & Brydon-Miller, M. L. (1984). Spatial and conceptual aspects of orientation: Visitor experiences at an outdoor history museum. *Journal of Environmental Education, 13*(4), 317–332.

Hayward, J., & Rabineau, P. (1989). The Pacific is too big: A problem of visitor orientation for a new exhibit. *Current trends in audience research.* Washington, DC: AAM Committee on Visitor Research and Evaluation, 11–22.

Heinrich, C., Appelbaum, K., & Birney, B. (1993). A formative evaluation of Habitat Africa's Thirsty Animal Trail. In D. Thompson, A. Benefield, S. Bitgood, H. Shettel, & R. Williams (Eds.), *Visitor studies: Theory, research, and practice, volume 5.* Jacksonville, AL: Center for Social Design, 148–162.

Hilke, D. D. (1988). Strategies for family learning. In S. Bitgood, J. Roper, & A. Benefield (Eds.), *Visitor studies—1988: Theory, research, and practice.* Jacksonville, AL: Center for Social Design, 120–134.

Hilke, D. D. (1989). Computer interactives: Beginning to access how they affect exhibition behavior. *Spectra, 15*(4), 1–2.

Hilke, D. D., & Balling, J. (1985). *The family as a learning system: An observational study of family behavior in an information rich environment.* (Final Report Grant No.: SED-8112927). Washington, DC: National Science Foundation.

Hilke, D., Hennings, E., Springuel, M. (1988). The impact of interactive computer software on visitors' experiences: A case study. *ILVS Review*, 1(1), 34–49.

Hirschi, K., & Screven, C. (1988). Effects of questions on visitor reading behavior. *ILVS Review*, 1(1), 50–61.

Hood, M. (1983). Staying away. Why people choose not to visit museums. *Museum News*, 61(4), 50–57.

ILVS Review: A Journal of Visitor Behavior. 1988–.

Jarrett, J. E. (1986). Learning from developmental testing of exhibits. *Curator*, 29(4), 295–306.

Kavanagh, G. (1992). Dreams and nightmares: science museum provision in Britian. In *Museums and the public understanding of science*, Durant, J. (Ed.), Science Museum, London.

Kimche, L. (1978). Science centers: A potential for learning. *Science*, 199, 270–273.

Koran, J., Koran, M. L., Foster, J., & Dierking, L. (1988). Using modeling to direct attention. *Curator*, 31(1), 36–42.

Koran, J., Koran, M. L., & Longino, S. (1986). The relationship of age, sex, attention, and holding power with two types of science exhibits. *Curator*, 29(3), 227–235.

Koran, J., Lehman, Shafer, & Koran, M. L. (1983). The relative effects of pre- and post-attention directing devices on learning from a 'walk through' museum exhibit. *Journal of Research in Science Teaching*, 20(4), 341–346.

Koran, M., & Koran, J. (1979). The study of aptitude-treatment interaction: Implications for educational communications and technology. *Instructional Communications and Technology Reports*, 9(3).

Korn, R. (1988). Interactive labels: A design solution. *ILVS Review: A Journal of Visitor Behavior*, 1(1), 108–109.

Korn, R. (1990). Men and women: Do they experience exhibits differently? In S. Bitgood, A. Benefield, and D. Patterson (Eds.), *Visitor studies: Theory, research, and practice, volume 3*. Jacksonville, AL: Center for Social Design, 256–262.

Lakota, R. (1975). *The National Museum of Natural History as a behavioral environment— Part I—Book I. (Final Report)*. Washington, DC: Smithsonian Institution, Office of Museum Programs.

Lakota, R. (1976). Good exhibits on purpose: Techniques to improve exhibit effectiveness. In *Communicating with the museum visitor: Guidelines for planning*. Toronto, ONT: Royal Ontario Museum.

Laetsch, W. (1991). Personal communication.

Laetsch, W., Diamond, J., Gottfried, J. L., & Rosenfeld, S. (1980). Children and family groups in science centers. *Science and Children*, Vol. 17(6), 14–17.

Loomis, R. (1987). *Museum visitor evaluation: New tool for management*. Nashville, TN: American Association for State and Local History.

Loomis, R. (1988). A note on the use of small samples to collect visitor information. *ILVS Review: A Journal of Visitor Behavior*, 1(1), 109–111.

McDermott-Lewis, M. (1990). *The Denver Art Museum Interpretive Project*. Denver, CO: Denver Art Museum.

McLean, K. (Ed.). (1991). *Recent & recommended: A museum exhibition bibliography with notes from the field*. Washington, DC: National Association for Museum Exhibition (NAME), American Association of Museums.

McLean, K. (1993). *Planning for people in museum exhibitions*. Washington, DC: Association of Science-Technology Centers.

McManus, P. (1985). Worksheet-induced behavior in the British Museum (Natural History). *Journal of Biological Education*, 19(3), 237–242.

McManus, P. (1987). Its the company you keep. . . . The social determination of learning-related behaviour in a science museum. *International Journal of Museum Management & Curatorship*, 6, 263–270.

McManus, P. (1988a). Do you get my meaning? Perception, ambiguity, and the museum visitor. *ILVS Review*, 1(1), 62–75.

McManus, P. (1988b). Good companions: More on the social determination of learning-related behavior in a science museum. *International Journal of Museum Management & Curatorship*, 7(1), 37–44.

McManus, P. (1989a). Oh, yes, they do: How museum visitors read labels and interact with exhibit texts. *Curator*, 32(3), 174–189.

McManus, P. (1989b). What people say and how they think in a science museum. In R. Miles & D. Uzzell (Eds.), *Heritage interpretation, Vol. 2: The visitor experience*. London: Belhaven Press, 156–165.

McManus, P. (1990). Watch your language! People do read labels. *ILVS Review*, 1(2), 125–127.

McManus, P. (1991a). Making sense of exhibits. In G. Cavenaugh (Ed.) *Museum languages, objects and text*. Leicester: Leicester University Press, 35–46.

McManus, P. (1991b). Towards understanding the needs of museum visitors. In G. D. Lord & B. Lord (Eds.), *The manual of museum planning*. London: H.M.S.O., 35–52.

McManus, P. (1993). Thinking about the visitor's thinking. In Bicknell, S., & Farmelo, G. (Eds.), *Museum visitor studies in the 90s*. London: Science Museum, 108–113.

McManus, P., & Miles, R. (1993). *Focusing on the market*. Museum International, 178(2), 26–32.

Melton, A. (1935). *Problems of installation in museums of art*. New Series No. 14. Washington, DC: American Association of Museums.

Melton, A., Feldman, N., & Mason, C. (1936). *Experimental studies of the education of children in a museum of science*. New Series No. 15. Washington, DC: American Association of Museums.

Miles, R. (1986). Lessons in "Human Biology:" Testing a theory of exhibition design. *International Journal of Museum Management and Curatorship*, 5, 227–240.

Miles, R. (1988). Exhibit evaluation in the British Museum (Natural History). *ILVS Review: A Journal of Visitor Behavior*, 1(1), 24–33.

Miles, R. (1993). Grasping the greased pig: Evaluation of educational exhibits. In Bicknell, S. and Farmelo, G. (Eds.), *Museum visitor studies in the 90s*. London: Science Museum, 24–33.

Miles, R. & Alt, M. (1979). British Museum (Natural History): A new approach to the visiting public. *Museums Journal*, 78(4), 158–162.

Miles, R., Alt, M., Gosling, D., Lewis, B., & Tout, A. (1988). *The design of educational exhibits (2nd ed.)*. London: Allen & Unwin.

Miles, R., & Tout, A. (1978). Human biology and the new exhibition scheme in the British Museum (Natural History). *Curator*, 21(1), 36–50.

Moscardo & Pearce (1986). Visitor centres and environmental interpretation: An exploration of the relationships among visitor enjoyment, understanding, and mindfulness. *Journal of Environmental Psychology*, 6, 89–108.

Nedzel, L. (1952). *The motivation and education of the general public through museum experiences*. Unpublished doctoral thesis. University of Chicago, Chicago, Illinois.

Norris, N. (1993). *"Bias and evaluation."* Paper presented at Museum visitor studies in the 90s conference, London.

Ogden, J., Lindburg, D., & Maple, T. (1993). The effects of ecologically-relevant sounds on zoo visitors. *Curator*, 36(2), 147–156.

Oppenheimer, F. (1986). Exhibit conception and design. In L. Klein (eds.), *Exhibits: Planning and design*. New York: Madison Square Press, 208–211.

Peart, R. (1984). Impact of exhibit type on knowledge gain, attitudes and behavior. *Curator*, 27, 220–227.

Porter, M. (1938). *The behavior of the average visitor in the Peabody Museum of Natural History*. New Series No. 16. Washington, DC: American Association of Museums.

Pratt Museum. (1992). *A summative evaluation of "Darkened waters: profile of an oil spill."* Unpublished report. Pratt Museum, Homer, Alaska.

Rabb, G. (1969). The unicorn experiment. *Curator*, 12(4), 257–262.

Raphling, B. & Serrell, B. (1993). Capturing affective learning. In *Current Trends in Audience Evaluation & Research, Volume 7*. Washington, DC: American Association of Museums Committee on Audience Research & Evaluation (CARE), 57–62.

Reque, B. & Wilson, L. (1979). *Getting the message out of the bottle . . . From Shedd Aquarium to Chicago schools.* Paper presented at American Association of Museums annual meeting.

Roberts, L. (1990). The elusive qualities of "affect." In Serrell, B. (Ed.) *What research says about learning in science museums*. Washington, DC: Association of Science-Technology Centers, 19–22.

Roberts, L. (1992). *From knowledge to narrative: Educators and the changing museum.* Unpublished doctoral thesis. University of Chicago, Chicago, Illinois.

Robinson, E. (1928). *The behavior of the museum visitor.* New Series No. 5. Washington, DC: American Association of Museums.

Robinson, E. (1930). Psychological problems of the science museum. *Museum News*, 8(5), 9–11.

Rosenfeld, S. (1979). The context of informal learning in zoos. *Journal of Museum Education: Roundtable Reports*, 4(2), 1–3, 15–16.

Schiele, B. (1992). Creative interaction of visitor and exhibition. In D. Thompson, A. Benefield, H. Shettel, and R. Williams (Eds.), *Visitor studies: Theory, research, and practice, volume 6*. Jacksonville, AL: Center for Social Design, 54–81.

Screven, C. G. (1973). Public access learning: Experimental studies in a public museum. In R. Ulrich, T. Stachnik, & J. Mabry (Eds.), *The control of human behavior, vol. 3*. Glenview, IL: Scott-Foresman, 226–233.

Screven, C. G. (1974). *The measurement and facilitation of learning in the museum environment: An experimental analysis.* Washington, DC: Smithsonian Press.

Screven, C. G. (1975). The effectiveness of guidance devices on visitor learning. *Curator*, 18(3), 219–243.

Screven, C. G. (1986). Exhibitions and information centers: Some principles and approaches. *Curator*, 29(2), 109–137.

Screven, C. G. (1990a). Computers in exhibit settings. In S. Bitgood, A. Benefield, & D. Patterson (Eds.), *Visitor studies: Theory, research, and practice, volume 3*. Jacksonville, AL: Center for Social Design, 130–138.

Screven, C. G. (1990b). Uses of evaluation before, during, and after exhibit design. *ILVS Review: A Journal of Visitor Behavior*, 1(2), 36–66.

Screven, C. G. (1992). Motivating visitors to read labels. *ILVS Review*, 2(2), 183–211.

Screven, C. G. (1993). *"Visitor studies: an introduction" and "United States: a science in the making."* Museum International 178, No. 2, 4–12.

Screven, C. G. (Ed.) (1993). *Visitor studies bibliography and abstracts, third edition.* Shorewood, WI: Exhibit Communications Research, Inc.

Serrell, B. (1981). Zoo label study at Brookfield Zoo. *International Zoo Yearbook,* Vol. 21, 54–61.

Serrell, B. (Ed.). (1990b). *What research says about learning in science museums.* Washington, DC: Association of Science-Technology Centers.

Serrell, B., & Raphling, B. (1992). Computers on the exhibit floor. *Curator,* 35(3), 181–189.

Serrell, B. (in press). *Why Ask Why.* American Association of Zoological Parks & Aquariums 1993 Annual Conference Proceedings.

Shettel, H. (1967). *Atoms in Action Demonstration Center impact studies: Dublin, Ireland and Ankara, Turkey.* (Report No. AIR-F58-11/67-FR). Washington, DC: American Institutes for Research.

Shettel, H. (1973). Exhibits: Art form or educational medium? *Museum News,* 52, 32–41.

Shettel, H. (1976). *An evaluation of visitor response to "Man in his environment."* Report No. AIR-43200-7/76-FR. Washington, DC: American Institutes for Research. Also Technical Report No. 90-10. Jacksonville, AL: Center for Social Design.

Shettel, H. (1989). Evaluation in museums: A short history of a short history. In D. Uzzell (Ed.), *Heritage interpretation, volume 2: The visitor experience.* London: Belhaven Press, 129–137.

Shettel, H. (1990). Research and evaluation: Two concepts or one? In S. Bitgood, A. Benefield, & D. Patterson (Eds.), *Visitor studies: Theory, research, and practice, volume 3.* Jacksonville, AL: Center for Social Design, 35–39.

Shettel, H., Butcher, M., Cotton, T., Northrup, J., & Slough, D. (1968). *Strategies for determining exhibit effectiveness.* Report No. AIR E-95-4/68-FR. Washington, DC: American Institutes for Research.

Shettel, H., & Schumacher, S. (1969). *Atoms in Action Demonstration Center impact studies: Caracas, Venezuela and Cordoba, Argentina.* (Report No. AIR-F58-3/69-FR). Washington, DC: American Institutes for Research.

Silverman, L. (1990). *Of us and other "things": The content and functions of talk by adult visitor pairs in an art and a history museum.* Unpublished doctoral thesis. University of Pennsylvania.

Sneider, C., Eason, L., & Friedman, A. (1979). Summative evaluation of a participatory science exhibit. *Science Education,* 63(1), 25–36.

St. John, M. & Perry, D. (1993). A framework for evaluation and research: science, infrastructure and relationships. In Bicknell, S. & Farmelo, G. (Eds.), *Museum visitor studies in the 90s.* London: Science Museum, 59–66.

Stevenson, J. (1992). The long-term impact of interactive exhibits. *International Journal of Science Education,* 13(5), 521–531.

Taylor, J. B. (1963). *Science on display: A study of the U.S. science exhibit—Seattle World's Fair, 1962.* Seattle: University of Washington, Institute for Sociological Research.

Taylor, S. (1986). *Understanding processes of informal education: A naturalistic study of visitors to a public aquarium.* Unpublished doctoral thesis, University of California, Berkeley, California.

Taylor, S. (Ed.) (1991). *Try it! Improving exhibits through formative evaluation.* New York: New York Hall of Science.

Uzzell, D. (1993). Contrasting psychological perspectives on exhibition evaluation. In

Bicknell, S. and Farmelo, G. (Eds.), *Museum visitor studies in the 90s*. London: Science Museum, 125–129.

Visitor Behavior: A publication for exhibition type facilities. Jacksonville, AL. 1986–.

Weiss, R., & Boutourline, S. (1963). The communication value of exhibits. *Museum News*, 42(3), 23–27.

White, J., & Barry, S. (1984). *Science education for families in informal learning settings: An evaluation of the HERPlab project*. Technical Report No. 86-45. Jacksonville, AL: Center for Social Design.

Wolf, L., & Smith, J. (1993). What makes museum labels legible? *Curator*, 36(2), 95–110.

Wolf, R., & Tymitz, B. (1979). *Do Giraffes Ever Sit? A study of visitor perceptions at the National Zoological Park*. Washington, DC: Smithsonian Institution, Office of Museum Programs.

Wolf, R., & Tymitz, B. (1981a). *The evolution of a Teaching Hall: You can lead a horse to water and you can help it drink*. (Special report, Office of Museum Programs, Smithsonian). Washington, DC: Office of Museum Programs, Smithsonian Institution.

Wolf, R., & Tymitz, B. (1981b). *"Hey Mom, That Exhibit's Alive": A study of visitor perceptions of the coral reef exhibit, National Museum of Natural History, Smithsonian Institution*. Washington, DC: Office of Museum Programs, Smithsonian Institution.

4

Evaluation in Informal Science Education: Community-Based Programs

Heather Johnston Nicholson, Ph.D.,
Faedra Lazar Weiss, M.A.H.L.
and Patricia B. Campbell, Ph.D.

> *This trip, the big discovery was two sting rays which had washed up on the beach. One was totally dried up and the other was soft. The girls experimented to see whether the hard one would become soft after soaking in water . . . many [of the girls] wanted to know more about the animal and how it had died . . . the girls were definitely thinking about biology and growth and death and decay. And they definitely got messy!! and took risks! (J. Martin, field notes, cited in Nicholson & Matyas, 1988)*

Informal Science in Community-Based Programs

The girls were members of the Girls Incorporated center in their working-class Massachusetts town and were on one of many visits to the beach as part of Operation SMART, a program to sustain girls' interest and persistence in *Science, Math and Relevant Technology* through inquiry, discovery, risking the untried, and getting wet and grimy in the process.

In the past, many community-based organizations have included math and science activities in their programming for youth in an unselfconscious way, along with sports, arts and crafts, and camping. Along with trips to zoos and science centers, reading *National Geographic*, and watching Mr. Wizard or Mr. I. Magination on television, these experiences were probably an important part of the "acculturation to science" (St. John & Shields, 1988) of today's scientifically literate adults, whether or not they earn a living in a scientifically-based field.

Today many organizations for youth and adults continue to offer their participants occasional excursions into science, mathematics, and technology. To set a context, in the United States some thirty million young people participate in the 15 largest youth-serving organizations (National Collaboration for Youth, 1990), with more than 17,000 organizations offering some community-based programming for young people (Carnegie Corporation of New York, 1992). There are no good statistics as to how many of these organizations offer programming in science-related areas, let alone in what degree of seriousness. The programs and organizations reviewed here are particularly serious about science—not surprisingly, since they have made the effort to assess their effectiveness in conveying science. Overwhelmingly, the evaluated community-based programs are designed for young people of school, and sometimes college, age. It is thus easy to presume that their goal is to supplement a more or less satisfactory system of formal education in science and mathematics. As will quickly become clear, to the contrary, many of the community-based programs were developed in reaction to inadequacies in the science curriculum and practices of schools. The programs are rich with insight and struggle about how to involve young people meaningfully and joyfully in the challenges of scientific and mathematical thinking. In this, they are far ahead of many school systems in implementing strategies recommended by experts in science and math teaching: for example, cooperative learning and hands-on experience leading to critical thinking and an ability to solve problems. The community-based programs thus provide important lessons about science education as well as about informal science and assessment.

The programs of "informal science education"[1] offered in community organizations, science centers, and the media have a number of characteristics in common:

- They are overwhelmingly voluntary. People "vote with their feet" whether or not to participate, so the programs must be attractive and exciting to survive.

---

[1] "Science," as explored in "informal science education," may include not only biology, chemistry, and physics but astronomy, electronics, the varieties of engineering, computer programming and applications, and other related topics, in any number and combination. Programs in "informal math education" cover an equally wide range of mathematical topics (symmetry, tessellation, measurement, estimation, probability and statistics, and logic, just to name a few); in addition, many "informal science education" programs include mathematics. Except where specific topics are noted to emphasize the focus of a given program, in this chapter we use "science" as shorthand for any or all topics in mathematics and science and "informal science education" for programs encouraging exploration of any or all of these areas.

- Participants generally are not tested on or held accountable for the science content they learn. No credentials need be issued for the participants to move to the next level of proficiency.
- They are designed more to generate interest and excitement than to provide a series of "right answers," to expand the pool of those who pay attention to science and who are willing to make the effort to understand the natural world.

Thus there are shared aspects of "informal" science education, as contrasted with the "formal" science education typically offered in schools and colleges, that make informal science difficult and elusive to evaluate. From the examples above, participants may not stay around long enough to be included in measures of success. It is difficult to design measures of what participants are taking home if it isn't concrete knowledge, and involvement in science may increase in difficult-to-capture minuscule increments or quantum leaps that may not occur immediately following an informal educational experience.

A thorough review of evaluation in informal science education should include specific attention to programs offered by community-based organizations. First, **innovative techniques of evaluation** have been used to assess the effectiveness of an array of exciting, **creative science programs.** Second, programs offered by community-based organizations tend to be **structured** differently than informal science education offered by science museums or media projects. Therefore, they present somewhat different challenges in evaluation. Third, the **purposes** for which many community organizations offer science programming are different from the purposes and perspectives of many museum and media projects. These different views raise important questions and issues about the nature of the scientific enterprise and who is legitimately involved in it.

Structure of Community-Based Programs

Blurred edges are part of any picture of informal science in community-based programs. When the defining characteristic is the structure of the programs, similar programs are offered by science museums, as part of media projects, and by colleges and university departments. When the defining characteristic is "out-of-school," programs developed and run by community-based organizations but offered during school hours are left out. When the defining characteristic is "informal," some exemplary discovery-based, process-oriented programs offered by schools are much closer to the definition than some of the most ambitious and carefully evaluated out-of-school programs. In this chapter, the focus will be on the programs that fall within the definition of **community-based pro-**

grams in **out-of-school contexts** using **informal approaches to science.** However, programs that violate one or more of these assumptions will be included, as they can greatly expand the foundation for understanding how programs can be evaluated and what works in community-based informal science programs.

In a pure form, an exhibit in a **science museum** exists in a fixed space and time. It is designed for people to interact with objects, to read some basic text or instructions, and to experience a connection with a scientific concept. It must often stand alone as a learning experience, attracting the participation of people widely different in age, scientific background, interest, and preparatory experience. Whether the exhibit is about mirrors in optics, the human gastrointestinal system, or the effects of pollutants on marine ecology, the encounter tends to be brief and often one-shot.

In a pure form, a **media project** goes through several stages in production, but once produced in final form it is relatively difficult to change or manipulate. It is designed to reach large numbers of people who nevertheless have great latitude in whether they participate and with what concentration. The typical media project engages primarily the senses of sight and hearing and lasts a fixed amount of time, either once or periodically. These characteristics are shared by a large-screen film on space travel, a half-hour television series to interest children in math, and a one-minute radio series on astronomy. Other media projects, such as multimedia science kits and specially designed computer software, are more flexible in structure but still often operate in such a way that the participant remains at a distance from the designer or purveyor of the experience. Like many museum exhibits, they must be self-contained and self-explanatory.

If there is a pure type of **community-based program,** it is targeted to a particular age group (usually youth), often with a shared socioeconomic background. The program emphasizes cooperative learning; participants interact with each other, as well as with materials from the natural world and perhaps equipment. It may last anywhere from a few hours to sustained participation over several weeks to several months or years. Finally, the making of conscious connections between the participant and careers in science and the relevance of science to daily life are almost universal goals. The program depends on adults, many of them not science professionals, and sometimes older youths to prepare materials and structure sessions. These characteristics are shared by after-school programs in science, math, pre-engineering and computers, summer math or science camps, day-long conferences on science careers for girls, and parent-child programs in math and science.

Formative Evaluation, Goals, and Outcomes

As will be discussed at greater length below, the developers and implementers of several community-based programs have been creative and thoughtful as they have designed and adapted programs, refined program goals, and created systems by which others could implement the programs. Often called **formative evaluation**, this consistent attention to observing how the program is going in order to improve it is the first obligation of program developers. It is often when programs are deemed by the creators, participants, and outside observers to be excellent and worthy of continued use in one place or even replication by other organizations that more stringent questions arise about "How do I know it's working?" Both the creators and potential implementing and funding organizations at that point ask whether the purposes have been successfully implemented, the goals are being met, or the program is having the **intended effects** upon the participants. Most of this chapter concerns the purposes of community-based programs in informal science and the search for these intended effects on participants, also called **outcomes.**

Community-based programs face numerous challenges to effective evaluation of outcomes. Even assuming would-be evaluators can keep track of participants long enough to involve them in evaluation, it is difficult to devise sufficiently sensitive instruments that the program can take credit for changes in knowledge, attitudes and behavior produced at low "dosage" or exposure to the program. Program effectiveness may depend on the efficacy of "pass-through" training of the adult or youth facilitators. The goals of the program do not always fit well with the ability to measure short-term and long-term impact. Always there is a tension in where time, energy, and other resources should be focused: on delivering the program, improving it, or demonstrating its effectiveness. Especially in informal programs, where participation is voluntary and many participants' perspective is recreational, the process of evaluation must be both brief and engaging for participants. The task is to find out whether the program is working without allowing the assessment tail to wag the program dog.

Purposes of Community-Based Programs in Science

Although there have always been stargazing in Girl Scouting and butterfly collections in 4-H, the **scientific** content or processes involved often have not been singled out as the primary reason for conducting these, rather than alternative, activities. Perhaps because scientific activities have been interspersed with others in programs designed to foster youth development, there was little impetus to assess their effectiveness as informal science education.

Experience

More recently, community-based organizations have been responding to perceived problems as they developed programs in science. Beginning in 1985 with a grant from the National Science Foundation (NSF), Hands On Science Outreach, Inc. developed voluntary weekly sessions of hands-on science in a recreational context for elementary school students. In part, developer Phyllis Katz and her colleagues were responding to the documented discrepancy between recommendations for hands-on, cooperative approaches to science from experts in science education on the one hand and the continuing pattern of lecture, vocabulary drill, and scant time on science at all that characterized elementary school science (Center for the Study of Testing, Evaluation, and Educational Policy [CSTEEP], 1992; Federal Coordinating Council for Science, Engineering, and Technology [FCCSET-CEHR PUNS], 1992; Gong, LeHart, and Courtney, 1991; Harty, Kloosterman, and Matkin, 1989). Children were not getting practical, hands-on experience with scientific materials and science concepts in school, so they needed this experience out of school and an organization was created expressly to provide the missing opportunity (Katz, 1992; Shettel, 1991).

Equity

Many other community-based programs were developed in response to concerns about equity in science. Both in the statistics about the scientific workforce (Massey, 1992; Zuckerman, 1987) and in portrayals in the popular culture (C.S. Davis, cited in George, 1990; Hardin & Dede, 1992; Kahle, 1987, and, for a historical look, LaFollette, 1992) science was a conspicuously white male domain. Yet high levels of poverty continued among Americans of color and among female heads of household. The discrepancy provided strong motivation for some groups to examine and attempt to redress the continued absence of underrepresented groups from the lucrative "scientific pipeline."

Continuing research indicated that there was no inherent intellectual reason for the underrepresentation in science of women, African Americans, Latinos/Latinas, Native Americans, and other groups; rather, the social forces perpetuating their exclusion were subtle, complex, and intransigent (Clewell & Anderson, 1991; Gibbons, 1992; Linn & Hyde, 1989; Oakes, Ormseth, Bell, & Camp, 1990; Zuckerman, 1987). Community-based organizations defined the challenge in a number of ways and devised a variety of strategies for increasing the participation of underrepresented groups in science, mathematics, engineering, technology, computer science, health professions, science teaching—the panoply of fields in which the numbers and proportions were low. The range of desired outcomes for these programs reflects an even greater range in the program developers' understanding of "the problem" and how to fix it.

Persistence

A fundamental aspect of the problem of equity in science is that a higher proportion of girls and women, and of students of color, than of white males drop out or fall out of math and science at all the critical junctures from elementary school through full professorship or corporate presidency. The presumption is that students of both genders and from all backgrounds are capable of doing the work in math and science and that community programs can provide some of the support and encouragement that are missing from the educational and larger social system as it exists. Some programs note the patronizing attitudes and low expectations female and minority students often encounter in schools, colleges, and the workplace. A number of strategies have been developed to challenge these assumptions and to encourage students' persistence in science in the formal educational system:

- combining tutoring and high expectations for sticking with tough courses;
- involving students in material more usually reserved for students in higher grades;
- involving freshman college students in support groups with classmates of the same gender, race, or ethnic background; and
- providing adult support in finding out about key points of decision affecting future education and career opportunities and negotiating the system to take advantage of these.

Countering the pattern of low expectations is only one of many strategies for which a positive outcome of a program would be the persistence of some or all of the participants in science.

Performance

Some programs emphasize not only continued participation in science, in or out of formal education (persistence), but also high achievement in science, math, and related subjects in formal education (performance). Here the measure of the success of a program is not whether the participants take a standardized test but whether they do well on it; not whether they enroll in a course but what their grades are.

Some of the programs that emphasize performance are more interested in whether participants have acquired skills of science as a process— comfort in approaching and manipulating unfamiliar materials, using observation to pose and answer questions, and understanding topics in deeper or more complex ways. In this view, the problem is that people who might be interested and capable in science are being driven out prematurely by pedagogy that does not resemble the real work of scientists and science workers or is not relevant or interesting to people from a particular culture (Duckworth, 1978; Heath, 1983; Trueba & Delgado-Gaitan, 1985). Program developers with these goals note that assessments conducted in formal education, many of them still focused on mastery of content, may be poor indicators of the success of a program that emphasizes science as a process or habit of thinking.

Improvement in Performance

Whether performance is defined as scores on standardized tests or ability to pose interesting questions, some programs emphasize high levels of performance for the "best" participants or for all of them, while other programs are more committed to producing change in the numbers or types of people performing well or adequately or improving the performance of everyone a little. This distinction between level of performance and improvement in performance often parallels the philosophy of a program about who can and should do science, as discussed in more detail below.

Love of Science

As some programs view it, getting people to stay in the scientific pipeline depends on providing them with enough positive opportunities to engage with science that they see science as interesting, relevant, fun, and a worthwhile way to invest time. Through discrimination, oversight, maldistribution of resources, and absence of opportunities and of role models, some young people more than others have been deprived of opportunities to see how wonderful science can be. The job of the program is thus to provide these opportunities; a program is successful if it increases the numbers or types of people who enjoy science or increases the level of enjoyment of the participants. For various people the love of science may be tied to the thrills of discovery and problem-solving, the intrinsic interest of a topic, appreciation for the aesthetic and physical dimensions of math or science, or pride in solving practical problems people face, just to name a few possibilities. The presumption is that there is a strong connection between liking science and engaging in it for fun or profit.

Confidence that One Can Do Science

To engage in science in school or as a profession, participants presumably need to see science as appropriate, accessible, and important to people like them. Some programs pay attention to providing participants with role models and mentors whose background characteristics (gender, racial and ethnic group, socioeconomic and geographic background, disability, family structure, . . .) are similar. This and other strategies are used to convey the message that **you** can and should engage in science, that science is not reserved for **others.**

Within this idea that science is appropriate and important for more people and more types of people than are currently pursuing it, there is nevertheless a tension about how open science should be, or where the emphasis should be in attracting more people. At the egalitarian end of the spectrum is the view that **everyone can and should do science.** For example, the American educational system is criticized for sorting students into those with and those without mathematics ability—often based more on stereotype than intellectual capacity. By contrast, other nations presume that all students have the capacity and the necessity to perform well in mathematics; the school system must and does produce that level of competence (Lapointe, Mead, & Phillips, 1989; Pechman, 1988).

In a similar vein, the gender stereotyping of the toys that adults give children—dolls and stuffed animals for girls, construction sets and transportation toys for boys—is an early part of a continuing pattern in which more boys than girls have experience in and encouragement for pursuing science and technology (Etaugh & Liss, 1992; Fry, 1990; Hardin & Dede, 1992; Nicholson, 1992). Thus some programs attempt to deliver the message that everyone, including **you,** can be good at science and related areas. These programs look for evidence that the participants are engaging in science with confidence and planning that they, their friends, and others like them will continue to pursue science. Many of the programs that emphasize confidence also work on the image of science and scientists to overcome the stereotypes that scientists are exclusively white, male, socially inept, strange-looking, nerdily dressed loners who work indoors with Frankenstein's monsters and things that explode (Kahle, 1989; Mason, Kahle, & Gardner, 1991).

At the elitist end of the spectrum is the view that science **requires special talent and extraordinary commitment** so that programs should identify, support, and prepare participants to join this special group. Programs with this view tend to recruit students already identified as academically talented or those identified as underachievers who are capable of advanced work. Presuming that previous tracking and sorting has underestimated the talents of underrepresented groups, some programs have chosen to work in formal education with whole classes to encourage new stars to emerge. Programs with an elite view have as a desired outcome participants' confidence that they have special aptitude for science and should pay attention to encouragement and overcome discouragement. The programs further look for evidence that the participants admire scientists and consider it socially acceptable for themselves and people like them to engage in science.

Progress Toward a Career in Science

When students leave the scientific pipeline prior to studying algebra and biology, it is difficult to know whether or not the world has lost Nobel laureates in physics or medicine. Community-based programs have taken a variety of perspectives on how important an outcome it is that participants intend, plan, take steps toward, or actually pursue a career in science. Developers of some programs primarily are interested in promoting scientific literacy and preparing the next generation of citizens for making personal and public choices that involve scientifically and technically complex issues (American Association for the Advancement of Science [AAAS], 1989; FCCSET-CEHR PUNS, 1992; Gilliom, Helgeson, & Zuga, 1992; Hazen & Trefil, 1990); if some participants increase their interest in a scientific career it is a salutary but not necessary outcome. For other programs, the expectation is that some of the participants will decide definitely on science, while others at least keep their options open. Some programs consider their efforts a failure if participants intend to pursue teaching, the arts, or law instead of science.

The range in what is considered **a career in science** as community-based programs define the term also is very broad. Often the programs designed to sustain the interest of girls and women focus on the lopsided gender distribution in the work force; the outcome is positive when girls and women pursue skilled crafts such as carpentry and plumbing or technical positions requiring a two-year degree or less in communications, electronic equipment repair, or the health or food industries. At the other end of the spectrum, some of the programs to encourage the participation in science of African-American students consider an outcome positive only if the participant goes on to complete a Ph.D. in a natural science and is appointed to a tenure-track position at a prominent university—construing the "scientific pipeline" in its narrowest sense. Similarly, pre-engineering programs may consider an outcome positive if the participant enters or completes college, or only if she or he completes a degree in engineering and is employed by a major engineering firm.

Reflecting on Purposes as Outcomes

Within the rubric of encouraging people to engage in science, the community-based organizations that create such programs differ significantly in how they define the problem and how to fix it. Some seek to expand scientific literacy as a societal characteristic. As seen above, many programs have a common purpose related to equity in science—to expanding the numbers and types of scientific insiders to include such underrepresented groups as women and people of color. Yet the strategies and emphases within this goal lead people to design quite different programs and to hope for quite different outcomes. The many challenges to evaluating the effectiveness of programs in producing their desired outcomes will be illustrated below, in the process of describing community-based programs for which outcome studies have been conducted.

One of the greatest challenges to evaluation is that most of the programs engage participants who are still in school, some as young as preschool age, yet the goals include changing the face of professional science and the scientifically literate citizenry. Program contact is brief and immediate but the goals are large and long-term. Many of the programs think about outcomes as short-term indicators for long-term goals. They thus have to guess which short-term, measurable outcomes are good predictors of longer-term goals. For example, in quantitative studies, does a seventh-grader's statement of her intention to sign up for geometry when she is in tenth grade predict that she will complete high school, go to college, take math or science in college, major in science, get a job that requires college math, enter graduate school, be employed as a scientist, or any of these—or even that she will sign up for geometry three years hence? In qualitative (naturalistic) studies, do seventh-grade girls who obviously enjoy doing science, ask pointed questions, and have cooperated to build and race a soap box derby car have a greater likelihood than girls who have not had such experiences to wind up as eager participants in policy-making about technologically complex issues?

The idea of outcomes in evaluations of community-based programs of informal science education is made more complicated by the need to make bets about both how to expand the group of people who engage in science and which short-term outcomes are good indicators of long-term goals. Add to this the challenge of documenting or measuring what the program is about and who attended it when the participants and often many staff members are volunteers and it is surprising that any examples of outcome studies are available to guide future designs for assessment.

Outcome Evaluation in Community-Based Programs

The number of outcome studies of informal science education in community-based programs in out-of-school contexts is fairly small. In this section the types of programs offered by community-based organizations are divided very roughly by structure and purpose into three groups:

Discovery programs—after-school programs with occasional participation in informal science and a focus on making engagement in science fun and rewarding

Science camps—intensive short-term programs, often residential or day camps, with course-like immersion in science and math and a focus on providing participants with the confidence and competence to pursue formal education in science

Career programs—longer-term programs with frequent participation and multiple strategies, including targeted contact with science professionals and a focus on preparing participants for careers in science

Clearly, these are not pure types. As a heuristic device, the division into discovery programs, science camps, and career programs permits an orderly presentation of the studies that come close to "**outcome** studies of **informal** science education offered by **community-based organizations** in **out-of-school** contexts," followed by discussion of similar programs with less evaluation or evaluation studies of programs farther from the pure type, including informal science in formal settings, to amplify how we know what works and how future evaluation studies might be conducted.

Discovery Programs

The **discovery programs** make science and math fascinating and exhilarating by having young people, sometimes with their parents, engage directly with the stuff of science and then reflect on their experiences. Science is something to **do,** not just something to **learn.** The premise in these programs is that **everyone** can and should do math and science; some of the programs have strong goals for promoting **equity** in science. Science and mathematics are portrayed as accessible, interesting, and useful pursuits, closely aligned with everyday occurrences. The focus is thus upon changing **attitudes** and inspiring **confidence** through positive experiences. Discovery programs are offered by youth organizations such as Girls Incorporated, Girl Scouts, and 4-H and by community organizations such as Hands On Science Outreach, deliberately created to provide opportunities to do science. One of the best-known programs was developed at a university but now is delivered either through schools or community organizations such as the National Urban Coalition and the Alba Society. The structure of most of these programs is like a club or a class—one signs up to participate and makes a commitment to appear one or more times a week for several weeks. The leaders act as facilitators, the activities include cooperative problem-solving and discussion, and an effort is made for the participants to form a cohesive and mutually supportive group.

Operation SMART and Gender Attitudes

Since 1985 the national youth organization Girls Incorporated, formerly Girls Clubs of America, has been developing and implementing Operation SMART, "a program to encourage every girl in science, math and relevant technology" (Girls Clubs of America [Girls Incorporated], 1986). The focus of Operation SMART is upon girls but many boys participate; the program's purpose is to empower young people who have traditionally been left out of science. Operation SMART is described as a web that makes connections for girls between science and the everyday world, a web spun of three E's—Exploration (scientific inquiry and a questioning, adventurous spirit), Equity, Empowerment—and one F—Fun (Girls Incorporated, 1990, pp. iii, 10).

With corporate, foundation, and NSF funding,[2] Operation SMART has

---

[2]Many of the informal science education programs described in this chapter have received NSF funding; many also have received funding from other foundations and/or corporations. Complete lists of funders for each program can be found in the references cited or by contacting the program office.

had the luxury of being created thoughtfully and deliberately, through collaborations and advisory councils, with the best experts on how a youth organization should do informal science. Although Girls Incorporated has written quite a bit about Operation SMART, including reports of ethnographic research and efforts to develop and test evaluation measures of attitude change, there has as yet been no coordinated national study of the outcomes for girls who participate in Operation SMART.

Beginning with a collaboration within Girls Incorporated of national staff and board with seven Girls Incorporated affiliates (then Girls Clubs) in the northeastern United States, Operation SMART now is implemented by more than 100 Girls Incorporated affiliates and by cooperating women's organizations, service clubs, and school systems nationwide. The program is replicated through intensive training (assisted by publications, training videos, and technical assistance) of staff members, most of whom are professional youth workers at Girls Incorporated centers and whose background in science ranges from very little through master's degrees and considerable teaching or research experience.

One of the Girls Incorporated affiliates that has implemented Operation SMART for a number of years is Girls Incorporated of Rapid City, part of a larger organization called Youth and Family Services. About 55 percent of the girls served by Girls Incorporated of Rapid City are Native American, primarily Lakota; most of the rest are white. Most participants are from low-income families, many of which are coping with drug and alcohol abuse. As part of a grant from the Bush Foundation to expand Operation SMART in South Dakota by training rural teachers to implement the program in their classrooms, the organization requested funds to conduct an outcome study of Operation SMART as implemented in the after-school program at the Girls Incorporated center in Rapid City.

The staff of Girls Incorporated of Rapid City worked with Steven Rogg, assistant professor of science education at the University of Maine, to develop a plan for outcome evaluation of Operation SMART, focusing on attitude change as a function of frequency of participation in the program. A major challenge for voluntary after-school programs is designing an evaluation that will allow them to know whether they can legitimately claim to have "produced" an outcome. Rogg designed a system in which a series of pencil and paper instruments were completed by girls every time a SMART group "turned over"—that is, when no girl who had previously completed the instruments participated in a given SMART session—or every ten weeks, whichever came sooner (Rogg, 1991). The system of course depended on very accurate records of participants' attendance.

The attitude instruments were adaptations by Rogg of brief, simple instruments deemed to be closely related to the objectives of Operation SMART as implemented in Rapid City. Four of these "tools" were adapted from the *Operation SMART Research Tool Kit* (Girls Clubs of America [Girls Incorporated], 1988), a set of fourteen activities for girls to measure their own and each others' attitudes toward math and science developed by the national organization of Girls Incorporated. The tools adapted were Draw-A-Scientist (DAST), based on the work of Chambers (1983) and Kahle (1989), in which participants draw a picture of a scientist at work and then select characteristics (male/female, long hair/short hair/bald, with something that explodes) that describe their picture; Math Attitudes Scale, adapted and condensed by Meyer and Koehler from the Fennema-Sherman Mathematics Attitudes Scales (MAS) (Fennema & Sherman, 1976); Science Attitudes Scale (SAS), adapted by Rogg from the MAS; and Survey, an exercise in which girls decide whether careers can be held by men, women, or both. A final instrument was Kahle's Science Experiences Survey (1983, cited in Rogg), used at the beginning of the evaluation to find that fewer than 20 percent of the participants had ever "worked with science related hobbies," "planted something and watched it grow," "found a fossil," or "cared for an unhealthy animal" (Rogg, p. 10). This was contrasted with the list of activities in which Operation SMART participants actually engaged, including investigating magnetism and conductivity, making rain and dew, making "fossils," making "volcanoes," growing beans, building ant colonies, building standing arches from blocks, and measuring trees by the stick method and with instruments (Rogg).

The SMART staff at Girls Incorporated administered the instruments and also maintained current information by name or unique identifier of age, grade in school, race, family socioeconomic status, dates of attendance at the Girls Incorporated center, and dates of attendance at Operation SMART. They also maintained session data, including a session evaluation form, equipment list, list of manipulative materials used, journal of relevant and/or exciting observations, list of activities including date, and objectives and a written description of each activity. Rogg made site visits to observe sessions in action and to interview participants and staff.

Rogg hypothesized a positive correlation between the number of SMART sessions attended and scores on the instruments measuring science and math attitudes and experiences, suggesting that even modest levels in this preliminary study would constitute evidence that the program is worth continuing and a motivation for further refinement of an evaluation protocol. Of 356 SMART participants, 81 (23 percent) attended one SMART session and 211 (59 percent) attended fewer than eight sessions. The mean number of sessions attended was eight, with 11 girls attending more than thirty sessions.

Rogg reported measurable positive correlations with all the instruments, though several relationships were not strong. On DAST, for example, a very high 37 percent of the participants drew a woman as a scientist, suggesting that the existence of Operation SMART at the Girls Incorporated center or previous participation in SMART (prior to the first "test") might be influencing the participants' expectations. As Rogg notes, previous research (e.g., Kahle, 1989) has found that children have very stereotypic views of scientists and that it is especially unusual for children to draw female scientists, although girls tend to do so more often than boys. Rogg did find some remaining stereotypical responses and recommends continued use of the activity, "especially as an activity which increases awareness of the nature of stereotypes. Also, results from the DAST should be used to guide programming. For example, these data suggest a need to convey to the girls that 'real' scientists are not necessarily male, unkempt, or 'nerds' " (p. 16).

Nicholson and Hamm (1990), working with Campbell, found that many girls in Girls Incorporated centers drew female scientists the second time they participated in DAST even without an intervening program or activity to encourage a reduction in stereotyping of scientists. They noted that "most of our early administrations suggest that this tool is one with troublesome confounding of the instrument with the intervention. . . . At least some positive change in attitudes occurs, as Pat Campbell puts it, 'without the fuss and bother of a program' " (p. 2). With Rogg, they recommend the activity as a tool for assessing the need to address stereotypes of scientists in programs for children.

Mason, Kahle, and Gardner (1991) used the DAST as a post-only measure of differences between experimental and control groups in a school-based program to encourage girls in science through positive attitudes and a broader, less stereotyped view of science and scientists. They found a strong difference in the predicted direction in the numbers of scientists depicted as female but no difference in the number of stereotypical characteristics depicted. They interviewed the students, designed a new rating mechanism, and trained new raters in its use. In the reanalyzed data they found that students in the experimental group were more likely to show scientists outside the laboratory and less likely to depict scientists as eccentric (nerds) or sinister (mad scientists).

DAST is a good example of the confounding inherent in trying to measure attitude change in the voluntary context. Nicholson, Hamm, and Campbell (Nicholson & Hamm, 1990) argue that evaluators need to search for tools that have sufficient test-retest reliability that they can be used twice in the absence of an intervention and found to change little. Then such reliable tools should be tested for "sandwich validity"—the ability to register change from before to after the "filling" of an intervention designed to foster a change in attitudes toward science. The purposes of needs assessment, programmatic intervention, and program evaluation are difficult to keep separate and the contribution of the program to attitude change is elusive when the change might come from the environment (for example, a Girls Incorporated center with a focus on girls' achievement in all areas, not just science), the program, some other source, or the instrument itself.

To return to Rogg's study, results on the MAS and the SAS were in the predicted direction but weak; they did not reach statistical significance.[3] Rogg's strongest finding of attitude change, accounting for 38 percent of the variance, was on the sex role survey. Participants in many SMART sessions were much more likely than participants in few sessions to say that both men and women (rather than only men) should have math- or science-related jobs such as airline pilot, auto mechanic, or bookkeeper.

In preliminary tests of the sex-role survey instrument Nicholson, Frederick, and Hamm (Girls Incorporated, 1992a) found that most older girls stated that both men and women should have these jobs; that is, there was a "ceiling effect" with the instrument prior to intervention. It may be that these girls, who were in Girls Incorporated centers, already had been "taught to the test" prior to their enrollment in Operation SMART or that expectations in the larger society have changed enough that girls, in some communities at least, will say the roles are open to both genders. The authors hypothesized that age was a critical variable, with older children cognitively more capable of distinguishing between possible and frequent patterns of occupation by gender; they thus designed the second version of the instrument for older girls (over age 9) to interview younger girls using the survey. The tension between the persistence of gender-role stereotyping in science on the one hand and the social unacceptability of espousing gender-role differences on the

---

[3]The MAS also has been used in evaluations of math camps. Further discussion of these scales will be reserved for the section on science camps.

other adds to the difficulty of assessing the effectiveness of programs in broadening the view of who can, does, and should do science.

Overall, the Rogg study of Operation SMART at Girls Incorporated of Rapid City is an ingenious design for trying to capture modest change over time in a program with shifting, voluntary enrollment. In principle, the structure of linear regression (positive outcomes increase as number of sessions increases) is elegant and helps to overcome the lack of a control group. In practice, his regression analyses were conducted with disturbingly small numbers and, as Rogg is the first to admit, with instruments that may or may not be able to detect change in attitudes toward science or voluntary engagement in science beyond the program.

Hands On Science Outreach, Inc. and Liking To Do Science

Another preliminary outcome study of an after-school, voluntary program is that of Hands On Science Outreach, Inc. (HOSO). The mission of HOSO is "to provide an enticing, regular structure for 'recreational practice time' for science-related activities from pre-kindergarten to sixth grade [and] to develop and disseminate such activities and materials, through community involvement, to as many children as possible through training of interested adults" (Katz, 1992, p. 1). Beginning in Montgomery County, Maryland in 1980, HOSO had delivered 11,833 classes representing 94,664 hours of programming in 30 states by the spring of 1992 (Katz). The program is delivered after school at school, once a week for an eight-week session, three sessions a year, by adults who are trained and paid to conduct it (Goodman, 1992; Katz).

Professional evaluator Irene F. Goodman of Sierra Research Associates prepared a report on the program Hands On Science (HOS) (1992) based on interviews with participating children and extensive process evaluation by Harris Shettel (1991). Questionnaire data from area coordinators trained by HOSO national staff and from area leaders trained by area coordinators indicate that the training is considered effective in preparing the adults to deliver the program with its philosophy intact and the sessions carried out generally as planned. Questionnaires from parents indicate that almost all children "often" or "sometimes" talk about what they did in HOS, bring projects home, and talk about these projects (Goodman; Shettel). The 10 percent response rate makes the parent data heuristic at best.

Goodman and her colleagues conducted a pilot study (1992) of outcomes by interviewing children in three age groups from 11 HOSO sites.

Efforts were made to interview each child three times: before the first session, at the fifth session, and at the eighth session. Due to dropouts and absences there were complete data for 34 children: 16 boys and 18 girls; 12 in kindergarten and first grade, 13 in the second and third grades and 9 in the fourth to sixth grades. The ethnic distribution was 52 percent Caucasian, 33 percent African-American, 17 percent Hispanic and 9 percent Asian, described as fairly representative of the children served by HOSO.

Nearly two-thirds of the children said they had taken Hands On Science before, and after the eighth session an average of 91 percent said they would tell a friend to take HOS (Goodman), evidence that the children enjoy their participation in the program. At the fifth session the two older groups were able to recall what they had done the previous week, with the fourth to sixth graders able to explain in detail: "For instance, the girl who said 'we worked with weather, made things to tell that hot air could rises, measuring wind speed and temperature' " (p. 23). Also at the fifth session the children were asked how HOS is different from school. The youngest participants thought HOS was more fun than school, the middle group said HOS was more fun because of their personal involvement ("Our other class we talk about stuff from a book—this class we *do* stuff" [p. 24]) and the oldest children said that, unlike science in school, in HOS they did get hands-on experience ("HOS is more interesting because in my regular room we don't do all these experiments" [p. 24]).

The only before-to-after questions in the interviews concerned the definition of science. To look for change over time the answers were coded into "don't know," "simple explanation," "fuller explanation," and "inquiry/experimentation." Goodman concludes that the younger group tended to move from "don't know" to "fuller explanation" and the two older groups from "simple explanations" to "fuller explanations" or "inquiry responses," giving examples of each (pp. 26, 28). However, in the accompanying bar graph (p. 27), no percentage change in the proportion of "inquiry" responses from the first to the eighth session is evident in any of the three groups. This may be a problem of capturing change in a coding scheme or of moving from coding to percentage tables when the numbers are small.

Goodman points out that *any* change toward an understanding of science as a process of inquiry and discovery must be considered quite positive. She cites Carey, Evans, Honda, Jay, and Unger (1989) and Grosslight, Unger, and Jay (1991) as finding that seventh-grade students (age 12) "regarded scientific knowledge as a passively acquired, faithful copy of the world" (Goodman, p. 28) and considered scientific inquiry to be limited to observing nature rather than constructing hypotheses about it. Grosslight and colleagues also confirmed these perceptions on the part of eleventh-grade honor students (Goodman, p. 28).

In general, Goodman concludes that the interviews were capable of detecting some modest growth over time and that some questions yielded valuable if preliminary information. Next time she would retain the questions "What is science?", "How is HOS different from science in school?", and "What was a new idea you learned?", using all these questions in the same form before and after the program. She recommends a participant observer over the eight weeks, who could ask questions on an ongoing basis, rather than the hurried interviews before class used in this pilot effort.[4]

Goodman was able to compare the interview process with focus groups of children at similar ages conducted by Shettel the previous year. Goodman and Shettel note that "focus groups with young children present difficulties because they suffer from 'bias and contamination'; consequently, they should be viewed with some caution" (Goodman, 1992, p. 29). Nevertheless, the findings from the interviews were consistent with those from the focus groups: "The fact that children in two different program years at different sites and in different contexts were able to talk about what they did at HOSO, said they enjoyed the program, and would recommend it, is further evidence that HOSO is meeting its major objectives" (p. 29).

FAMILY MATH and the Home Environment

FAMILY MATH (FM) is a program developed at the Lawrence Hall of Science at the University of California, Berkeley, and disseminated across the country in relationships with community organizations. In FAMILY MATH, parents interact with their children in engaging, hands-on activities designed to make mathematics fun, exciting, understandable, and related to the real world. The program is designed to increase the quantity and quality of parents' interaction with their children around math at home as a way to support the children's

---

[4]As this volume was going to press, we received a new evaluation of HOSO (Goodman & Rylander, 1993). This study, which includes observation of participants and interviews of participants and a comparison group, confirms many positive outcomes of program participation while noting some persistence of the idea that science is only concrete and product-oriented.

persistence and performance in math at school (EQUALS, 1992; Weisbaum, 1990).

Kathryn Sloane Weisbaum of the Lawrence Hall of Science conducted "a qualitative study of parents' roles in their children's formal (school) and informal (out-of-school) mathematics learning" (abstract). The conceptual model that served as a basis for the study predicts that increases in parents' own knowledge and self-confidence about mathematics would lead to increases in the number and type of family activities and resources related to mathematics, leading to increased opportunities for the child to learn skills relevant to mathematics, resulting in increased achievement in mathematics. The study examined the impact of change in one component—parents' knowledge and beliefs—on other aspects of the home environment.

The research staff worked with the FAMILY MATH site leaders at each of four sites: local schools in the San Francisco Bay area (Caucasian families), the National Urban Coalition in Washington, D.C. (African-American families), the Alba Society in San Diego (Mexican-American families) and the Oregon Indian Society in the Portland, Oregon area. Through a subcontract, leaders at each site recruited and hired two local interviewers. Interviewers were of the same ethnic background as the families to be interviewed and had some experience with FM but were not directly affiliated with the classes being conducted at the site. All interviewers at each site participated in two days of training, including practice with the tape recorders they would use. They also helped to review and amend the protocols (recruitment materials, consent forms, interview schedules) to make the materials less intimidating and more friendly for the particular groups of parents. The documents and interview protocols were translated into Spanish for the Mexican-American sample.

In order to recruit families in all sites but Portland, where the program was undergoing reorganization, the interviewers visited the first session of a FAMILY MATH group, usually of seven to fifteen families, and asked parents to fill out a statement of interest. Oregonian participants were recruited from rosters of FM classes. The plan was to include two or three families from each FM group; more expressed interest than could be interviewed. All parents were contacted but those selected provided a balance in age (grades 3 to 5) and sex of the children attending. "We also wanted to interview fathers whenever possible" (Weisbaum, p. 5).

The final groups included 15 families from each of the four ethnic groups. Eighteen percent of the sample were fathers; in the total sample most of the parents were married and lived with their spouses. Most of the mothers had finished high school with some additional courses in junior college or college, with more of the Mexican-American mothers not having completed high school and more Caucasian mothers having finished college. Fathers living in the home were about equally likely to hold professional or office jobs and skilled or technical jobs, with a higher proportion of incomplete data about Mexican-American fathers.

The first interviews were conducted between the first and second FM sessions, at a date, time, and place arranged between interviewer and parent(s), and lasted 60 to 90 minutes. The same interviewer contacted the parents at the end of the series of FM classes to conduct the post-interview within two weeks and the follow-up interview four months later. (Oregonian families could be interviewed only once, retrospectively.) Interviewers in the other three sites also visited the FM classes that included parents on their case list to observe the program, establish a common ground for later discussions, and interact informally to establish rapport. The observation notes and any other interviewer's notes on interactions with the parents or children became part of the "case," which also included the typed transcripts of the three interviews and the interviewer's notes on the context of the interviews (Weisbaum).[5]

As a measure to investigate self-selection bias in the parents who were interviewed, a survey, "loosely derived" from the Mathematics Attitudes Scales (Weisbaum, p. 6), was administered when possible to all the parents in the FM class. The survey responses of those interviewed were very similar to those of the remaining parents, suggesting little self-selection bias. Interestingly, most of the survey scores were extremely positive. Though this might be expected from parents who took the trouble to sign up for a family experience in mathematics to support their children, parents' responses in interviews were not so positive or consistent, leading Weisbaum to doubt the validity of the survey.

The data were analyzed by looking for "themes, patterns, commonalities and distinctions among the cases" (Weisbaum, p. 7). All the first interviews were searched for common patterns of parental attitudes, beliefs, and behaviors; all the post- and follow-up interviews for parental reactions to the FM experience and ways parents incorporated FM ideas and activities at home. Cases within each group were compared and contrasted with an eye toward "perspectives, concerns, attitudes or situations" (p. 7) unique to each of the four groups. A formal report for

---

[5]This combination of observing the program sessions and arranging interviews so that they are not rushed by the proximity to the program is what Goodman seems to be recommending in her remarks on the preliminary study of HOSO (1990).

each group was written by a staff member who specialized in that specific cultural group.

For the sample as a whole, the parents' beginning beliefs about mathematics constitute some of the most interesting findings of the study:

> The parents in our sample were committed to supporting their children's schooling and placed a high value on achievement in mathematics as a key to future career opportunities. But parents' own formal educational experiences had led them to view mathematics as a difficult and inaccessible school-based subject, largely unrelated to practical applications in the "real world." Their views of how mathematical concepts and skills are taught and learned were based on traditional approaches that relied on an expert demonstrating the "one right" algorithm, followed by repetitive drill and practice to memorize the correct procedures. *Given these views of mathematics, parents saw their own roles in the development of children's mathematical competencies as limited to monitoring the completion of homework and checking homework papers for neatness and correctness. They were careful not to demonstrate different ways to solve problems lest they confuse the child. Their efforts to teach their children "everyday" skills in mathematics were not seen as relevant to the child's school performance, nor were the parents generally aware of ways in which they might be supplementing or extending the child's learning through informal activities and interaction* (Weisbaum, abstract; italics added).

Those parents interviewed overwhelmingly viewed participation in FAMILY MATH positively; only four or five said it was "boring" or "not worthwhile." Again, Weisbaum's own language gives a succinct statement of the direction and intensity of some of the changes the parents reported, often attributed to their FM experience:

> Given their own prior experience in mathematics classes, many parents approached FAMILY MATH classes with trepidation. They found, however, that the instructional approach presented in FM represented a dramatic shift from the traditional approaches these parents remembered from their own schooling. For many parents, FM had a profound influence on their perceptions of mathematics as a subject (math can be fun and very creative), of how children can learn mathematical concepts and skills (through games, hands-on activities, and problem-solving challenges) and of the pervasiveness of mathematics throughout their daily lives. Most gratifying to many parents was the opportu-

nity to work with their child on intellectual activities not related to homework. Through these interactions, parents gained new insights into their children's learning styles and mathematical competencies. Armed with these new perspectives and insights, parents began experimenting with different techniques for helping with homework and for expanding their children's awareness of the ways in which mathematics pervades their lives. *While only a few parents reported frequent use of FM activities at home several months later, many more were drawing upon FM ideas and strategies in helping with homework and in emphasizing math skills in family games and interactions* (Weisbaum, abstract; italics added).

Although outcomes for children were not directly assessed in the study, Weisbaum notes that parents generally thought their children seemed to like mathematics more, and to have more positive attitudes about it at school and at home. Some parents noticed improvement in their children's math grades; most did not expect participation in Family Math to affect math grades, at least in the short run. Nearly all parents said they would recommend FM to other parents.

The African-American parents from Washington, D.C. seemed especially concerned that their children develop positive attitudes toward school in the face of a high dropout rate. Caucasian parents seemed more concerned with perseverance and achievement. Some African-American and some Caucasian parents were actively involved in the alternative public schools their children attended, paying close attention to school work and developing strong, positive relationships with school personnel.

The FAMILY MATH experience was reported to be especially intense for Mexican-American families from the San Francisco Bay area, interviewed in pilot tests of the protocol. All were recent immigrants, many were non-English-speaking and had little formal schooling. While many of the Mexican-American mothers in the San Diego area actively volunteered at their children's schools, the Bay area parents had counted on mathematics, less subject to the language barrier, to be able to connect to their children's schoolwork but found the mathematics education system in the United States very different from their own. For these parents FM was an important opportunity to interact with their children in a school-related activity.

Similarly, the Native American families from Oregon "felt isolated and threatened by the dominant culture represented by the school system . . . the school represents a threat to their culture and an obstacle in the way of their children's chances for a happy and successful life" (Weisbaum, p. 29). FM helped these families make connections between valued traditional arts and school mathematics and science.

Informal Science as Discovery

Operation SMART, Hands On Science Outreach, Inc., and FAMILY MATH, as reflected in the outcome studies, have in common a perspective that children who participate in science and mathematics in their most interesting forms are more likely to persist in these subjects. There is an implicit or explicit distinction between "real" science (hands-on construction of explanations for fascinating, frequently encountered natural phenomena—presumably the sort of activity professional scientists are paid to engage in) and "school" science (the memorization and recitation of "right answers" to arcane questions of interest only to the ancient or modern nerds who came up with them). The reasoning, apparently supported by the studies, is that children are interested in science and mathematics; they are turned off by the tiresome pedagogy of science and mathematics that occurs in most schools most of the time. Thus, these informal programs are not merely supplementing what happens in school; they are competing with school for the opportunity to define science in order to retain enthusiasm for engaging in it. If school science is as deadly as portrayed, these programs may also be raising the number of experiences of science-and-math-as-discovery by a factor of n in the lives of participants. It then becomes especially important to know not only whether the programs produce persistence as an outcome, but also how and why they achieve this end.

It is an empirical question whether it is "school" science rather than "real" science that leads to an exodus from science, especially of girls and of students of color and students with disabilities. Much of the literature cited near the beginning of the chapter (CSTEEP, 1992; FCCSET-CEHR PUNS, 1992; Gong et al., 1991; Harty et al., 1989) indicates that the negative characterization of "school" science is not far off. However, an ethnographic study of messages to girls about science in three Girls Incorporated communities (Frederick & Nicholson, 1991) suggests that we may be scapegoating schools by oversimplifying the conflict as between school science—that is, science taught in schools—and informal science. These authors found that schools, Girls Incorporated centers, and other institutions within a given community gave girls quite similar messages. Conversely, individual communities differed considerably in the messages and opportunities they provided girls to engage in science-as-discovery, to take risks and pursue a question tenaciously, to cooperate in solving problems, to make big, interesting mistakes in pursuit of understanding the natural world. They reported that in all three communities studied, despite the differences from one community to the next, girls faced the barriers of "underexposure and underexpectation" to their continued involvement in science. Similarly, Weisbaum (1990) found that parents in her interviews prior to FAMILY MATH participation not only seemed to have the "debilitating" attitudes towards interest and competence in science and math produced by negative school experiences, as suggested by the work of Stodolsky (1985) and Schoenfeld (1987), but were in subtle ways reinforcing these attitudes in their children.

As Weisbaum notes, Resnick (1987) draws the distinction between "practical, real-world intelligence and learning" and "school intelligence and learning"[6] as follows:

- individual cognition in school versus shared cognition outside of school;
- pure mentation in school versus tool manipulation outside of school;
- symbol manipulation in school versus contextualized reasoning outside of school; and
- generalized learning in school versus situation-specific competencies outside of school.

Operation SMART, Hands On Science Outreach, and FAMILY MATH clearly are consciously striving to purvey **practical intelligence.**

In the three programs described above, the outcomes assessed were immediate to short-term. Indeed, although Weisbaum conducted interviews six months after the intervention, the analysis combined the interview data from the post- and follow-up interviews. The outcome measures, whether quantitative, as in Rogg's study of Operation SMART, quantified interviews, as in Goodman's study of Hands On Science Outreach, or qualitative, as in Weisbaum's study of FAMILY MATH, searched for change in attitudes toward science at least in part through a revised conception of science-as-discovery.

Where a set of surveys and scales that have **sandwich validity** can be found, the quantitative study has a number of advantages. Surveys are easy to administer and interpret, impose little on the participants or the

---

[6]At the risk of belaboring the point, not all "school science" is formal, dry, dusty, or irrelevant and not all "out of school science" is informal, practical, cooperative, or enjoyable; see the descriptions of the African Primary Science Project and the "schooly" EUREKA! teacher for counterexamples.

program time, can be used by a wide variety of programs with similar goals, and if used by many programs could begin the development of a national database of comparative data. Interviews are more expensive, require training of interviewers not directly involved in the program, depend on the outside time of participants, and are considered by some researchers to be less **objective** measures of change.

However, especially when supplemented with observation of the program in action, interviews have the distinct advantage of connecting the change in knowledge, attitudes, and behavior directly to the intervention. Students who can contrast school science with Hands On Science and parents who can contrast their and their children's school math with FAMILY MATH have been affected by the intervention. Moreover, the combination of interviews and observations reflects whether the intended intervention, allowing for local adaptation, actually has been delivered. The reader has confidence that FAMILY MATH had traveled through training and technical assistance to the four sites and perhaps further as site directors trained the particular teachers whose students were interviewed. The characterizations of and direct quotations from parents did indicate that the FM experience was both quite similar and quite true to the spirit of the program for participants from the four sites and ethnic groups.

Other Examples of Out-of-School Science Discovery

CTW and Multimedia Kits. Laura Martin, Carol Ascher, and Mark St. John (1991) conducted studies of multimedia kits produced by Children's Television Workshop (CTW) around the television series *3-2-1 Contact* (a science program) and *Square One TV* (emphasizing math concepts) as used in school-age child care programs operated by YMCA, Boys and Girls Clubs, and other organizations across the country. Although they were seeking to refine the kits rather than to measure educational outcomes for children, they found that their intensive study, which included interviews and on-site observation, provided important information for making the resources work in context. The multimedia kits were being used in situations, unlike those of the discovery programs above, in which "the idea of doing math and science is a foreign one to the after school culture" (p. 9). The researchers found that the *Square One TV* materials tended to be used to fill an existing "games niche"; the *3-2-1 Contact* materials, an "arts and crafts" or "literacy" niche. Although the kits were used to foster what Resnick would call practical knowledge, often by children and adults engaging in the process of trying to make the activities work, the resulting learning was not closely associated with traditional notions of science and mathematics problem-solving. They concluded that, for the intervention to carry from one setting to the next, the kits would need to provide numerous choices, include activities ranging from basic to sophisticated and usable by individuals and groups over a range of ages, and fit into existing after-school programming "niches."

From the perspective of outcome studies, the Martin, Ascher, and St. John study illustrates two important points:

- Even well-packaged programs are not self-executing. The outcome study must provide a mechanism for insuring that the program as delivered is faithful to the original goals or to some revised set of goals that are still worth pursuing.
- There may be no substitute for observation and interview to establish what, in fact, the intervention is.

AAAS and Girl Scouts: Linkages for the Future. The American Association for the Advancement of Science (AAAS) and 14 Girl Scout councils in the upper Midwest combined their expertise to develop Linkages for the Future. Under the direction of Marsha Lakes Matyas, this program provided comprehensive manuals and equipment kits for trainers and troop leaders, but depended on carefully designed training of adults to disseminate the intervention. Matyas has designed several survey instruments to measure a number of outcomes for girls, including participation in the activities and attitudes toward math and science, while encouraging councils to collect information on courses girls take in middle school. Data collection with these instruments is continuing within the 14 councils as Matyas continues to refine the instruments (AAAS, 1992; M.L. Matyas, personal communication, December 2, 1992).

One ingeniously unobtrusive measure of outcomes is the number of badges related to science and math awarded to girls. Badges are awarded when girls demonstrate that they have had experiences, mastered skills, understood concepts, built or made things, and so on, and thus are evidence of specific outcomes. They are available only through official Girl Scout channels, so that the numbers can be tracked efficiently by date, place, and in many cases by purchasing troop. Leaders provided data on the number of badges in science and math their troops worked on, showing a 150 percent increase after program implementation;[7]

---

[7]Matyas confirmed that these data were collected before revisions in the badge programs in 1989 and 1990; the current availability of new and revised badges thus is not a confounding factor (personal communication, December 2, 1992).

numbers of badges sold by the councils provide an independent check on these figures.

As other researchers, Matyas was unwilling to measure the outcome of the program for girls without first establishing whether the program had taken place. Thus she began with extensive data collection on whether the pass-through training system was functioning in the numbers of training sessions held, the numbers of leaders trained in the participating councils, the numbers of sessions and activities leaders actually conducted with girls, and so on, as well as whether the trainers and leaders felt confident of their ability and preparation for conducting activities (AAAS, 1992; *"Linkages for the Future,"* 1992).

4-H SERIES. The 4-H SERIES (Science Experiences and Resources for Informal Educational Settings) program in California is tinkering with participant **portfolios** as a way to document both the character of the intervention and the outcomes. In SERIES, teen leaders are trained to lead science activities for younger students, ages 9–12, to increase the science literacy of both groups. On such engaging topics as Recycle/Reuse ("Will a glass bottle still be around in the year 4990?"), Beyond Duck and Cover ("Can a 3-story tower built on Jello withstand an earthquake?") and It Came from Planted Earth (What kind of corn grows in a bucket of water?) teens lead younger children through activities designed to teach scientific processes (observing, communicating, comparing, ordering, categorizing, relating, inferring, and applying). To solidify their knowledge, participants engage in a community service project of their own design (4-H SERIES Project, 1992; R.C. Ponzio, personal communication, November 9, 1992).

4-H SERIES project director and principal investigator Richard Ponzio is developing a system for measuring the outcomes of the program for the teen leaders (4-H SERIES Project; Ponzio). The leaders keep journals of their activities, maintain attendance records and anecdotes of the sessions they run for younger children, and document the community service projects. They may collect quotations, drawings, or charts from the younger participants, photographs of events, reflections on their own training experience or consultation with adults and other teen leaders. The teen leaders select and assemble these materials literally into a portfolio that can be reviewed by others. Ponzio plans to have members of an advisory board or another group of professionals familiar with the program and its objectives review the portfolios, comparing the science knowledge, attitudes, and behavior of the teen participants with that of other teens in their experience.

The **portfolio** strategy has the advantages of being integral to the program and thus unobtrusive, relatively inexpensive, and a good source of information about the intervention and immediate outcomes. A further challenge will be to design the portfolio plus a review system to reflect *change* in the participants' knowledge, attitudes, and behavior (as a result of the program) as distinct from the post-program *status* of what was learned.

Assessing Science Learning

A few techniques for assessing the outcomes of discovery science are worthy of special note, although to our knowledge they have been used only in school settings. As with extensive interviewing and observation, these techniques are more likely to be used in a research study of outcomes than as a part of routine documentation and evaluation by program developers and implementers. Certainly they are appropriate only in the more intensive programs described in this section, not after a few semistructured opportunities to engage in hands-on science.

Observing Diversity and Complexity: African Primary Science Program. In the African Primary Science Program (Duckworth, 1978), elementary students were taught in a manner consistent with discovery learning, with special emphasis on presenting interesting "stuff" which children then explored, manipulated, and asked questions of. When children asked, "What does an ant lion eat?" the reply was "Ask the ant lion; it will give you the answer" (Duckworth, p. 43, citing African Primary Science Program, 1969), and children went back to observe the ant lion and offer it different foods.

When children were thoroughly familiar with some sets of manipulatives and with the discovery process of learning, Duckworth implemented a system of observation to see whether the "diversity and complexity" of these students' interaction with scientific materials was systematically different from that of students who had been in more traditional science classes. The system has the advantage of producing scores for individual students, providing statistical power in comparing one group with another. Twelve students from a class are randomly selected to be studied. They wear identification numbers front and back while they are provided as a group with unfamiliar sets of scientific materials as manipulatives. Each student in turn is observed by each of the two adult observers six times during the approximately half-hour session. The first and last observation rounds allot ten seconds per student and the intervening rounds thirty seconds, for a total of nearly five minutes of observation per student. Student activities were rated for type (off task, watching another child, handling or sorting materials, examining a single object), complexity (simple, moderate, elaborate, and extraordinary), and diversity (variety of uses of the same or different materials). For example, test materials included equipment for electrical circuits, including a simple circuit set up, and Playplax, interlocking translucent plastic pieces. Simple to moderate uses of these materials included copying the model circuit less or more successfully and using the Playplax pieces as color filters; in an extraordinary activity, one student used these materials to build a house with illuminated rooms. In Duckworth's original study the differences between the discovery classes and the regular classes in diversity and complexity of activities were strong and statistically significant at $p \leq .01$.

Assessment of Performance Unit's System. The Duckworth system for observing and recording children's comfort and sophistication with hands-on science is more useful in the context of community organizations than some other systems that may be more widely known. In the Assessment of Performance Unit (APU) system developed in England, for example, children are presented not with materials but with hands-on tasks that they complete and then write about. The tasks have a "right answer" that the child is supposed to find and report, even though the focus ostensibly is on science as a process. Patricia Murphy, an expert on the APU science tests, reports (1989) that prior experience out of school confounds some of these measurements, with effects along gender lines. At ages 13 and 15 "boys' performance is better than girls' precisely on those instruments that they claim to have more experience of outside school" (p. 326). She also notes performance effects arising from the "combination of avoidance by some pupils and the heightened confidence of others" (p. 327), as seen by girls' higher scores on such content as health, nutrition, reproduction, and domestic situations and boys' higher scores on building sites, racing cars, and electricity.

The APU, at least according to the current scoring systems, may perpetuate precisely some of the screening factors that select for already competent and experienced boys over girls and those with less experience in science:

This problem[8] concerned human survival when stranded up a mountainside. The material was for making a jacket. The actual investiga-

---

[8]See Murphy and Gott (1984), pp. 9–13 and 43–45, for the problem ("Survival") posed as a hands-on experiment and the checklist used to rate investigations.

tion was heavily cued to be concerned only with thermal conduction and pupils were even recommended to use cans of hot water and to lag them with the materials provided. However, for some of the girls the context overrode the cued investigation and they stayed with the broader problem presented, in this case, as a motivating context. Thus, it mattered how porous the material was to wind, how waterproof, and whether indeed it was suitable for making a jacket. There are at least three issues arising from this small scenario. First, if we intend to locate science problems within everyday or social settings then a broader view of what is relevant and of the range of problems that can legitimately arise needs to be understood, particularly if we wish to foster a divergent and critical approach to problem-solving. Secondly, if we change the values that underpin assessment it is important to recognise the effects of changing them in a short time-scale. At the moment such change would disadvantage some boys. Finally, while all teachers need to focus onto specific learning opportunities the route taken to this must take account of pupils' various and different values, experiences and expectations (Murphy, p. 330).

Thus the systems of observation that depend on prior restriction of the problem to be solved, that require individual performance when the programs encourage cooperation, or that sort participants finely by level of skill or competence, are of little use in the informal setting. To the contrary, the question is whether the *program* is working by helping some or all participants to increase their skill or enjoyment in problem-solving. Such concepts as the **complexity** and **diversity** of engagement with materials seem well matched to the ideal of discovery science.

Audiotutorials and Concept Maps. Novak and Musonda (1991) devised **concept maps** constructed from interviews with children who had or had not participated in audiotutorials about science as first- and second-graders. They sought "not only to explore what information might be remembered from the audiotutorial lessons but also to explore the child's ability . . . to transfer and apply concepts and propositions to explain novel phenomena" (p. 124). They taught graduate students to conduct noncuing, probing interviews, finding it was essential for these students also to have a firm grasp of the science concepts involved in the interview. The interviews were conducted periodically with the same respondents from grades one to twelve. The analysis of concept maps was done from verbatim transcripts of the interviews, sometimes resorting to tones of voice on audio recordings.

"The construction of concept maps permitted us to begin with the most general, most inclusive concept dealt with in the interview and to show propositional structures in a hierarchical arrangement, also illustrating important interrelationships among concepts included in different interviewee statements" (Novak & Musonda, p. 126). A first draft was revised two or three times by careful return to the written or taped interview. Trained and experienced research assistants, working independently, constructed maps that were "remarkably similar in the knowledge structure shown" (p. 126). A scoring system was developed, giving, for example, more points for acquisition of an alternative but closely related concept than to new examples of the same concept. Novak and Musonda found significant differences between the scores on concept maps of instructed and uninstructed students, both in sophistication and levels of misconceptions, favoring the instructed group as long as eleven years after program participation.

Few informal science programs may be ready for such sophisticated measures of science learning as Duckworth's system of observation or Novak's concept mapping. But these techniques seem closer in structure and goals to informal science than are most of the standardized tests and assessment systems used in formal education.

Science Camps

Community organizations, colleges and universities, businesses, and government agencies offer young people intensive experiences in math and science. These residential or day camp programs generally last two weeks or longer, usually in the summer. Many of the camps recruit participants who are capable of good work in science, math, computers, or pre-engineering but who are at risk of falling out of the pipeline. The approach to science is likely to be hands-on and informal but the sessions are likely to be structured, cumulative, and taught by people who earn their living in formal education. The message is less "science or math is play" than "science or math is work but you can be good at it and enjoy it." A socially and intellectually satisfying intensive encounter with these subjects, the reasoning goes, will increase participants' confidence that they can succeed in these subjects at school. If a high proportion of the other participants, leaders, and scientists encountered on field trips are perceived to be **like them** in socially significant ways, and the content is presented as consistent with rather than contrary to their culture, the reasoning continues, the participants are especially likely to learn that they can legitimately see themselves as scientific insiders rather than outsiders.

AISES Summer Math Program and Changing Math Attitudes

In the search for pure types of science camps offered by community organizations and with an available report on an outcome evaluation, the one that may come closest is a two-week residential program created by the American Indian Science and Engineering Society (AISES) and conducted on a small college campus in the Midwest. Authors John J. Hoover and Cathy Abeita, both associated with AISES at the University of Colorado, report (1992) that the participants were 25 rising eighth-grade students (14 female, 11 male) with grade averages of 3.0 or higher, representing 12 tribes and 11 states. During the two-week residential experience the participants engaged in courses such as probability and statistics, astronomy, computers, wide world of mathematics, and confidence; laboratory experiences in physics, geology, chemistry, and biology; field trips to the Milwaukee Museum, Crane Planetarium, and the Oneida Nation Museum and other trips involving compass orientation and making dream catchers; films including "Dances with Wolves" and "Stand and Deliver" followed by discussions; and guest presentations, including the Science of Alcohol Curriculum for American Indians, decision-making, and science fair ideas. Mathematics was carefully integrated into all the content areas and class sizes were small, 7 to 10 participants.

In addition to a rating of the program activities, participants completed four of the 12-item subscales of the Mathematics Attitudes Scales: attitudes toward success in mathematics, confidence in learning mathematics, usefulness of mathematics, and anxiety toward mathematics, comprising a 48-item survey. Hoover and Abeita conducted pre/post comparisons of means for the total group and separately for males and females, using correlated t-tests. They found significant increases in confidence in learning mathematics and in overall positive attitudes toward mathematics for the total group and a significant decrease in anxiety toward mathematics for the total group. The gender breakdown indicated that much of the change in the total group was attributable to change among the females, although the attitudes among males were generally positive and changed, though not significantly, in the predicted direction. Hoover and Abeita interpret the results to suggest that "even short-term programs which involve hands-on classroom and laboratory experiences, along with positive role models, affect anxiety and attitudes, particularly in female students" (pp. 5–6). They note the limitation of the short-term assessment (before and after 14 days of intensive intervention); they intend to continue to collect data from the same students, who are scheduled for campus and summer programs for three consecutive summers.

A little further from the pure type is EUREKA!, an intensive day camp for girls based at the Women's Center at Brooklyn College, a four-year coeducational liberal arts college. The AISES and Brooklyn College programs are included in part because community organizations conduct such programs but may not have conducted outcome evaluations and in part because they are important additions to what is known about outcome evaluations of programs with the **science camp** format and goals. Indeed, a project funded through a grant from NSF will test the Brooklyn College model at five sites through a Girls Incorporated adaptation of EUREKA!, as a precursor to national replication.

The EUREKA! Experience and Persisting Through High School

For eight years EUREKA! has targeted rising eighth- and ninth-grade Brooklyn girls from a variety of racial and ethnic backgrounds. This program is designed to help girls to capture the "Eureka!" ("I have found it!") feeling through individual and group achievement in mathematics, science, and sports. Participants first attend Math EUREKA! and have the option of returning the next summer for Science EUREKA! Both programs run from 9:00 A.M. to 5:00 P.M. five days a week for four weeks during the summer; follow-up sessions are offered on Saturdays during the spring.

Evaluation has been an important component of EUREKA! since its inception. Within the constraints of a small project budget and a minuscule evaluation budget, formative and summative evaluations have been conducted each year. All students complete questionnaires and interviews before and after the summer program in which they participate and at the end of the follow-up sessions in spring. In 1991, a longer-term follow-up was conducted, in which 1987 and 1988 EUREKA! participants were interviewed about their high school careers and their experiences after high school (Clewell, 1992).

The primary result of the evaluations is that for the most part, girls who participate in EUREKA! increase the numbers of math and science courses they plan to take. After participating in EUREKA! girls also were more involved in sports, were more interested in careers in engineering, evinced more self-confidence in math and science and in sports, and knew of more women involved in sports and in science (Campbell &

Shackford, 1990; Miller, 1990). Both the short-term and longer-term follow-up indicated that girls did continue their increased involvement and interest in these areas. The longer-term follow-up of participants in the first two years of EUREKA! found EUREKA! participants more likely than their peers in comparable national samples to take four years of high school math and science, enroll in college, and participate in sports (Clewell).

According to the participants, EUREKA! programs succeed because they present math and science as fun, hands-on, relaxed, and informal. This does not mean that EUREKA! is easy but rather that participants enjoy a well-designed challenge. Indeed, girls have commented that "EUREKA! math makes you use your brain; school math you just have to fill out blanks" and that "In school math is dull; all it requires is sitting at a desk. EUREKA! math requires a lot of skill" (Campbell & Shackford, 1992).

The greatest strength of the ongoing evaluation is the ability to ask the same questions over a period of years. Data on participants' continuing involvement in math, science, and sports and repeated measures of knowledge and attitude document the continuing impact of program participation. The consistent pattern of results does much to increase confidence in the accuracy of the results and suggests where to search for reasons when the pattern is broken. For example, girls participating in Math EUREKA! one particular summer did not increase the number of math courses they planned to take. The lack of change was traced to a math teacher who, while very well liked, was very "schooly" in her work with the participants. The math curriculum was modified accordingly and participants during the following summer returned to the pattern of increased math course taking (P.B. Campbell, personal communication, January 25, 1993).

The consistent ongoing evaluation has also increased flexibility in evaluation. After a period of consistent results, it may not be necessary to continue to test some variables. For example, after several years in which program participation produced no significant change in attitudes as measured by the Mathematics Attitudes Scales (MAS), use of these scales was discontinued after discussion with Dr. Fennema.[9] Participants' pre-

---

[9]The continuing saga of the Fennema-Sherman Mathematics Attitudes Scales (1976) has been a theme throughout the chapter. This tool seemed capable of measuring positive change in Operation SMART in Rapid City and in the intensive math program of the American Indian Science and Engineering Society. It lacked robustness in other Operation SMART settings, seemed not related to math attitudes as reflected in interviews in FAMILY

and post-program listings of notable women in sports and in science were dropped as well when the evaluators felt that there was no more to be learned from this measure. New emphases in the evaluation include more open-ended responses, allowing exploration of the "why?" behind changes in participant attitude, skills, or behavior. Thus EUREKA! participants are now asked to complete such sentences as:

I feel good when I. . . .
When I do math (science) I feel. . . .
When I do sports I feel. . . .
Girls who do math are. . . .

While EUREKA! has been able to implement and use a strong evaluation component, it suffers from many of the same problems related to evaluation that are found in other informal science programs. The primary problem is of course the lack of resources. Annual evaluation budgets for EUREKA! have ranged from $1,000 to $4,000; much of the work has been done *pro bono.*

A second major problem is the lack of useful reliable and valid measures. Many of the available attitude measures have never been tested for reliability. As noted earlier, just participating in some measures, such as drawing a scientist or listing science careers and indicating the ones you might do, produces a strong positive effect without the fuss and bother of a program encouraging changes in attitude. The few existing measures that are known to be both reliable and valid, such as the MAS, are designed to tap underlying personality constructs that are not easily changed so that an effective program may produce little change in response.

The use of consistent ongoing evaluation does, however, present the danger of a false sense of security. Changes in the program over a period of years, combined with an increasing tendency for participants to come into the program already more interested in mathematics and science, make cross-year comparisons difficult (Miller, 1990). Continuing evaluation of EUREKA! includes the development of analysis procedures for working with the larger quantity of open-ended data. Plans also are currently underway to combine the data collected from the annual evaluations with the data collected from the longer term follow-up and to do more comprehensive cross-year analysis (Campbell, 1992).

MATH in four communities, and failed to detect change among Brooklyn girls in EUREKA!. This may in part be due to the selection of different subscales in different circumstances but aspects of its usefulness as a tool for outcome evaluation remain a puzzle.

Other Science Camps—SummerMath, Summerscience,
and MISS

EUREKA! is not the only summer camp-like program to target girls and
young women. Among the first and most long-lived is SummerMath at
Mount Holyoke College in Massachusetts (Mount Holyoke College, 1991).
A program to sustain girls' interest in science is Summerscience, a
project of the Women in Science program of the Center for Continuing
Education for Women at the University of Michigan (George, 1990;
Nelson, 1990). In this two-week residential camp, some 75 rising ninth
graders from Michigan spend mornings doing field work in their chosen
specialty of space science, engineering, physics, chemistry, or natural
resources and their afternoons using computers to run experiments,
graph results, and write reports. Evenings include discussions of ethical
issues and learning about scientific careers. Director Cinda Sue Davis
reports that an outcome evaluation has been conducted of Summerscience and that results will be reported within the next year or so (C.S.
Davis, personal communication, December 15, 1992).

The Mathematics Intensive Summer Session (MISS) is a month-long
residential program of California State University, Fullerton. Involving a
racially and ethnically diverse group of young women who were rising
high school juniors (considerably older than the middle school/junior
high students targeted in many programs), the program's long-term goal
is to retain females in calculus to increase their options for college majors
in the sciences. Reports from 1990 and 1991 indicate gains in scores on
mathematics diagnostic tests and in positive attitudes toward mathematics, both for new groups and for a small number of members of the 1990
group who returned in 1991 (Pagni & Long, 1993).

In all these programs—the AISES summer program, EUREKA!, SummerMath, Summerscience, and MISS—the most critical outcome is
whether the participants enroll in college preparatory mathematics and
science in the formal educational system. Although increases in scores
on achievement tests or in grades are considered good news, the more
important factor is whether participants continue to behave as scientific
insiders—evidence of their confidence in math and science—when the
program is over. The outcome measures (e.g., course enrollment, a
variable for which the evidence indicates self-report is reliable [Crockett,
Schulenburg, & Petersen, 1987]) are relatively straightforward, with the
tracking of participants over a period of years the most cumbersome and
expensive requirement. So far, the results from the few intensive summer
programs that have follow-up data are encouraging, but much more

needs to be learned about the critical aspects of the various programs with remarkably similar goals.

The TERC Environmental Network Project and Science Mastery

The evaluation of the TERC Environmental Network Project (Spitzer & Foster, 1991) focused largely on the participants' mastery of scientific processes and concepts as principal outcomes, at least in this report on the pilot stage of implementation. TERC is a nonprofit organization commissioned by the Connecticut Department of Environmental Protection (DEP) to design, implement, and test the Environmental Network, an intensive science experience for 12- to 14-year-olds involving "in-depth science content, access to technology tools for data collection and analysis, and the use of telecommunications networks to link students with each other and with scientists and to access current scientific data" (p. iii). The project topic was air pollution, with a focus on the deleterious effects of ground-level ozone.

The summer program was piloted with 28 students in three "informal education settings": a science museum, a science center, and a nature center. Several techniques were used in this combined formative and summative study: questionnaires and interviews with teachers and students; observations, including videotaped sessions; and embedded performance assessments. Staff at the program sites reviewed the materials and decided on the duration and schedule of the Environmental Network Project: 2 hours a day for two 5-day weeks (20 hours) at the science museum, 4.5 hours a day for one week (22 hours) at the science center, and 1.5 hours a day for two weeks (15 hours) at the nature center.

Throughout the program students worked cooperatively and did quite a bit of analysis and reporting. In one of these embedded measures, about halfway through the program the students held a scientific conference on the nature and extent of ozone in their area and then wrote air quality data reports and recommendations for the Commissioner and for the other sites. The evaluation report's authors, William Spitzer and June Foster, both of TERC, interpret participant reports as indicating that the students "clearly grasped the relationship between ozone levels and higher temperatures/sunnier days" (p. 19). This was the concept most clearly understood and also the concept on which the students got the most direct practice: indeed, they conducted their own measurements, created their own data sets, analyzed and reported the data, and discussed the relationships between their data and the data collected by the other sites and DEP.

Overall, the findings were that students grasped the science concepts, were engaged in the curriculum, and "comprehended, and were keenly invested in, the policy and individual action implications of their findings about air pollution" (Spitzer & Foster, p. 39). With certain exceptions, participants' level of conceptual analysis and data analysis skills were reported to correlate highly with age, with younger students (some as young as 10, younger than the curriculum was designed for) grasping elementary concepts and skills and the students in grades 7–9 operating at the "intermediate" level. Few students mastered the most sophisticated concepts (the chemical processes that form emissions and the photochemical reactions that produce ozone from its precursors); Spitzer and Foster speculate that these concepts might be within the grasp of students in grades 10–12 who had had more exposure to chemistry.

The combination of teacher observations, embedded assessments in which the students were writing about their findings, and transcripts from videotaped sessions in which students interact with each other and the computer-based data analyses, proved to be important techniques for inferring the skill and understanding of the *group*, though not necessarily of individuals. On the one hand, these techniques for group-based assessment are entirely appropriate; the issue is whether the *program* produces desired outcomes for at least some students and not the relative achievement of individual students. On the other hand, the TERC evaluation illustrates both a potential dilemma of relying on group-based assessment and the special value of videotaped observations. The evaluators concluded that age was an important variable and reported no conclusions based on gender or racial-ethnic group. Yet in one of the three groups, two of ten participants were female and in another seven of ten were female. In one group half the participants (5 of 10) were described as minority and in another group all (8 of 8) were described as minority. For the developers of many summer intensive experiences, this natural laboratory of variation in the gender and racial/ethnic mix would provide opportunities to notice whether there is a "critical mass" of girls necessary for girls to feel comfortable doing science, whether minority students are half the leaders when they are half the students, whether white males accounted for a disproportionate share of the limited hands-on computer time, and what, if any, teacher and student strategies affected these patterns. A substantial videotape library of program sessions would make retrospective study of such questions possible. The point is not that Spitzer and Foster should have undertaken an analysis of gender and racial/ethnic equity—their analysis already is remarkably comprehensive for a program in its pilot stage—but rather that evaluations based on observation, when they exist only in real time, may miss collecting data on topics of vital interest to other analysts.

Intensive Experience Through Science Camps

The summer seems a natural time to boost the confidence of young people that they can and should engage in science. Whether offered by community-based organizations, colleges and universities, or professional associations, the preliminary evidence is that the strategy is effective in fostering **confidence** (AISES, EUREKA!) leading to **persistence** in science as reflected in subsequent school enrollments (EUREKA!, MISS). The strategy seems especially well suited to convincing members of underrepresented groups—Native Americans in the AISES project; New York City girls of color in EUREKA!—that they are potential scientific insiders. They spend two, four, or more weeks with adults and other youths who are like them both in background characteristics and in enthusiasm for exploring the natural world. The camping format gives program developers the freedom to address specific factors that may contribute to confidence and persistence, such as sports to encourage risk-taking by young women in EUREKA! and science as expressed in Native American life in the AISES program.

Preliminary studies also suggest that summer is a natural time to increase young people's knowledge of science (TERC) and mathematics (MISS) as measured by embedded assessments of concept mastery and tests of math proficiency. Whether focused on confidence and persistence or direct knowledge, intensive experience through summer camp almost certainly increases the participants' engagement with science by an appreciable factor over what most of them get through school.

Career Programs

If *discovery science* programs emphasize enjoyment and a process orientation that contrasts with the way science often is presented at school, and the *science camps* provide a one-time or repeated booster to confidence and experience leading to persistence in science and math in formal education, then the *career programs* tend to be multifaceted support systems designed to insure that the participants remain in the path toward a designated job or career. These programs are not so much criticisms of formal education or alternatives to it but mechanisms for seeing that students stay in the scientific pipeline despite barriers. The discovery programs vary widely in the amount and duration of any one person's participation, with an emphasis on the voluntary character of participating, while the science camps require a firm commitment to an intensive exposure to science and the program's values. The pure type of career program is **extensive,** with participation at least once a week for a year or more, often with a presumption of continuing involvement through college and beyond.

Project Interface, Performance, and the Pipeline

A program that establishes the pure type is Project Interface, a program begun jointly in 1982 by the Allen Temple Baptist Church in Oakland, California and the Northern California Council of Black Professional Engineers (Clewell, Anderson, & Thorpe, 1992).[10] Developed originally with a grant from the U.S. Department of Education, Project Interface now is supported by grants and contributions from corporations, foundations, and individuals.

The program recruits seventh, eighth, and ninth graders who are capable of doing college preparatory work but who are not enrolled in college prep courses. These students spend after-school periods four or more days a week in project activities:

- the math strand on Mondays and Wednesdays, to ensure mastery of current classroom topics;
- the science strand on Tuesdays, with fourteen weeks each in chemistry, physics, and biology, to spark interest and introduce fundamental concepts;
- a role model and mentor strand, involving professionals in math, science, and engineering who visit two or three times a month to demonstrate there are "paths to success other than crime, entertainment and athletics," (Clewell et al., p. 268, quoting the project director); or
- a career exploration strand, involving repeated field trips at two local high technology companies;
- a computer literacy strand at three different levels on Thursdays and Saturdays.

---

[10]The following description of Project Interface draws heavily upon the excellent compendium of formal and informal programs *Breaking the Barriers: Helping Female and Minority Students Succeed in Math and Science* (Clewell, Anderson, & Thorpe, 1992). With the permission of Project Interface directors, data reports were furnished by Beatriz Chu Clewell for this chapter.

The program also requires parental involvement and support, with parents monitoring attendance, participating in academic and career counseling, receiving monthly written feedback, and, along with the students, signing an annual contract committing themselves to the program's goals. Other features of the program include cooperative trips and projects with other area programs in science and math; motivational, academic, and career counseling; and incentive awards for attendance, achievement, effort, and excellence.

The math and science strands are planned and implemented by young college students, originally junior college students being groomed for transfer to four-year institutions and now involving a mix of two- and four-year college students. These tutors meet weekly with project staff for planning and review, meet twice a month with professionals to discuss their own prospects, and maintain careful records of the progress of the younger students they work with. They are "treated as professionals" and data on the college students' own grades and academic and career progress are maintained by Project Interface.

Project Interface has assembled and reported several types of outcome measures, reporting some of them annually from 1982–83 through 1986–87. Change scores on the California Test of Basic Skills (CTBS) were compared with the change scores of students from area schools and from the district as a whole. The project demonstrated impressive, sometimes stunning, increases compared to the non-program groups, measured as changes in the median score, changes in the grade equivalent or the number of months gained in grade level, and even in changes in the percentile rank on this nationally normed test. Thus, on this standardized test of performance, the scores of participants in tutoring and other strategies were more likely to increase than the scores of students not similarly engaged (Project Interface, 1987b).

Project Interface also reported on the enrollment of participants in college preparatory mathematics and whether the enrollment was at the expected level (on track) for their age and grade—for example, algebra as ninth graders and geometry as tenth graders. For the total group in 1986–87, 76 percent (48 students) were enrolled in a college preparatory math class. Of these, half were on track (Project Interface, 1987a). In a telephone interview study of ninth graders who had "graduated" from the program, with an impressive 81 percent response rate, a combined 85 percent of the graduates were enrolled in college preparatory math classes. Among those reporting for each year of graduates, 80 percent or more were earning a "C" or better and 40 to 60 percent were earning a "B" or better (Project Interface, 1987c). Although these data were not directly compared with data from nonprogram groups, the size and consistency of these groups of African-American students doing well in college preparatory math in the Oakland school system seems important evidence of the efficacy of Project Interface.

Clewell, Anderson, and Thorpe report that about half the junior college tutors in each of three reporting years transferred to four-year programs; several have gone on to graduate schools and prestigious positions (p. 275). Overall, through maintenance and follow-up of some basic records, some obtained from the school system and others by self-report of the participants and tutors, Project Interface has been able to monitor and demonstrate the achievement of basic project goals on an ongoing basis.

MESA and Institutionalizing Data Collection

A petroleum engineer from the University of California, Berkeley and a science teacher from Oakland Technical High School combined forces in 1969 to address the problem the teacher, Mary Perry Smith, described as follows: "If teachers spent as much extra time and energy on promising math and science students as the school coaches spend on promising athletes, the schools of Oakland would produce as many engineers as professional athletes" (Gibbons, 1992, p. 1191). The program they founded as the Mathematics, Engineering, Science Achievement Program (MESA) has expanded from 25 students at Oakland Technical High School to 14,000 California students in grades three through college. Wilbur Somerton, the petroleum engineer, answered the question "What makes MESA work?" by saying, "It is simple, really. You encourage students, you back them up, but you demand excellence" (p. 1191).

To serve this number of students MESA has been institutionalized in California. The program links universities, usually engineering schools, with schools and school systems and with professionals in industry. The statewide MESA system is organized into components by the level of education of the participants and the specific objectives parallel the objectives of maintaining high expectations and support for "widening the pipeline" to retain Mexican-American, African-American, Native American, and other students of color (California Postsecondary Education Commission [CPEC], 1992; Clewell et al., 1992). CPEC reports that in 1989–90 high school-aged MESA participants were over twice as likely to *complete* advanced mathematics and chemistry courses and over four times as likely to *complete* physics courses as their nonparticipating peers were to *enroll* in these courses (p. 40). MESA participants overwhelmingly live up to their aspirations to go to college; 85 percent of

1990 MESA graduates enrolled in postsecondary programs in California, with 74 percent in four-year colleges (CPEC, p. 43). Gibbons adds that "the MESA Minority Engineering Program (MEP) is responsible for two-thirds of the [engineering] bachelor's degrees awarded to blacks, Hispanics, and American Indians in California—producing a total of 600 engineers last June" (p. 1191).

The elaborate system of tracking and documentation that can report data at this level of sophistication comes from several influences: the career (as distinct from discovery or confidence) outcome goals of the program, access to school and college records through direct collaboration with these institutions, and the large and expanding scale of MESA, making the investment in the creation and maintenance of a database cost-effective (Clewell et al. 1992, p. 201).

Science Skills Center and Overcoming Underexpectation

The Science Skills Center in Brooklyn now recruits students in grades 2 through 6 from across New York City to become part of a "science club," a group of committed students who devote substantial time after school, on Saturdays and during the summer to discovering science as a way of understanding the world. The Center was formed in 1979 by African-American science professionals and teachers to support minority students who were doing well in school and were interested in science.

> Teachers' low expectation of students, many inexperienced teachers, and a lack of materials, positive role models, and cultural support from other students and adults combine to create a situation in which a student's success depends more on luck than pluck. In other words, a child's success may depend largely on having a good teacher in a class that is held to at least modest expectations and includes supplementary parental education, which in many cases is also essential (Johnson, 1991, p. 267).

According to Michael Johnson, Director of the Science Skills Center, the center began in 1988 to search for an assessment tool that would adequately measure all the skills the center teaches, reflecting the advanced capabilities of the third and fourth graders in the robotics program and the third through sixth graders in the life sciences program. After careful reflection on the advantages and disadvantages, the center staff decided to allow the center participants to prepare for the tenth-grade New York Regents examination in biology. The decision was motivated both by "the desire to compel the public education system to raise the level of performance expected from minority students" and to "empower students educationally" through the prospect of success in the examinations (Johnson, p. 269).

A number of educational challenges were confronted directly. A deliberate effort was made to prepare the students psychologically: "students who have been chronically under-challenged must be convinced that they can rise to a difficult occasion and succeed" (Johnson, p. 269). The strategies included:

- giving students a personal mission that puts them in control of the challenge;
- developing a sense of history and purpose and a sense of representing the achievements and aspirations of those to whom one is culturally linked;
- instilling a sense of team spirit; and
- creating a strong sense of ability that meets others' doubt with increased determination.

Second, the students had to be taught to retain and integrate information on diverse topics over a sustained period of time—quite different from the system many elementary students encounter of learning and being tested on a single topic or block of information.

Third, the students had to learn to decode the multiple choice questions, adopting a "bicultural approach" to understand what the examiner wanted in order to get credit for their knowledge of biology. "The students were taught how to dissect multiple-choice questions, filter out unnecessary data, analyze the remaining data (the givens), time themselves, relax before an examination, and, finally, check their answers after completing the test" (Johnson, p. 272).

As the students were studying for two hours after school and four hours on Saturday, they used the same textbook many high school students used. Many had to learn still another lesson in functioning in two cultures: to avoid embarrassing their teachers and getting into trouble in the regular classroom. Center students agreed "not to read in class, not to talk about the center at their '9-to-3' schools, and to give what they termed 'baby answers' to science questions in their regular classrooms" (Johnson, p. 271).

Johnson reports that in May 1989, 16 students in grades 4 to 7 took and passed the tenth-grade New York States Regents examination in biology. Johnson reports that these students were the youngest ever to have passed the test but he does not report the proportion of tenth-graders who take the test and pass. He argues that "the results of accelerated science courses and examinations are good ways of evaluating minority programs for high academic achievers." He argues further that the Science Skills Center demonstrates the failure of the public school system to establish an environment of high academic expectation in which all teachers and all students are excited about learning. Johnson argues in effect that the pattern of low expectation of minority students puts them in double jeopardy—first by underestimating the capability of all students of color and second by leaving the students who have special interest and potential with no system of support.

The approach of the Science Skills Center is ingenious in turning the hazards of "teaching to the test" into a positive benefit for the students who participate, preparing them for the existing system while providing an alternative, supportive environment. By focusing on students with interest and potential in science and then presenting a rigorous, intensive program, the Science Skills Center seems to be operating at the elite end of the continuum in preparing students to enter the scientific pipeline. The evidence that fourth graders can pass a test given to tenth graders is persuasive on content mastery, psychological preparation, motivation, and test-taking skill. Ideally, the evidence of short-term prowess would be followed up by documenting that center students continue to participate in science at high levels through high school, college, and beyond.

Other Career Programs

A variety of programs across the country are operated by or with the cooperation of such professional associations as the Association of Women in Science (AWIS) (C. Didion, personal communication, December 15, 1992), Women and Mathematics (WAM) (Kelly, 1991) and the American Chemical Society (ACS) (Carpenter, 1992). With their beginnings in community organizations, many of these programs work directly with the schools and colleges and so from the perspective of this chapter more closely resemble efforts in formal education. Some of the evaluated programs include clearly informal strategies and thus potentially important lessons in evaluating out-of-school programs, although the intervention often takes place at school during school time.

Project SEED and Extensive Intervention

Actually two projects, and perhaps many more, have been called Project SEED. The American Chemical Society has sponsored Project SEED as the Summer Educational Experience for the Disadvantaged. The Project SEED being discussed here was developed in 1963 by mathematician, psychologist, and teacher William Johntz in Berkeley, California. This Project SEED (Special Elementary Education for the Disadvantaged) trains industry mathematicians, university professors, and other subject matter experts, initially all community-based volunteers but now including some paid staff, to teach high school or college level mathematics to whole classes of students in elementary or middle school grades, using Socratic teaching styles and participatory learning-and-response techniques (Clewell et al., 1992; Webster & Chadbourn, 1992). "Project SEED believes that only persons who understand mathematics in depth possess the versatility to capitalize on the unconventional and often original insights that children are capable of making in an open-ended mathematical dialogue" (Webster & Chadbourn, p. i). The program is designed to foster skill and confidence for success in mathematics. Project SEED now operates in many school districts and serves more than 6,000 minority students in three school districts in California and in Philadelphia, Dallas, and Detroit (Clewell et al.).

Evaluation of Project SEED as it operates in the Dallas Independent School District is reported by evaluators William J. Webster and Russell A. Chadbourn. In Dallas the evaluated program was given one semester a year in grades four, five, and six, in supplementary classes (that is, not as a replacement for regular mathematics instruction) and to students of all levels (that is, not only disadvantaged students). The study, built on four studies previously undertaken of Project SEED in Dallas, involved elementary and middle school students, all of whom were black or Hispanic or both.

A series of sophisticated evaluation studies compared students who participated in Project SEED with "theoretical comparison groups" from the school district. That is, each student in the Project SEED classes was matched on sex, ethnicity, grade, socioeconomic status as indicated by eligibility for free lunch, and math achievement (total math score on the Iowa Test of Basic Skills [ITBS]) to another student somewhere in the school district. Thus, SEED was being compared to a variety of other mathematics contexts, although a consistent difference was that SEED students had an extra 45 minutes of mathematics a day four days a week for one semester a year (Webster & Chadbourn). Although school systems have distinct advantages over community organizations in setting up the structure of a nearly experimental evaluation, the idea of a "theoretical control group" might be more manageable for the community organization than, say, random assignment to treatment and control groups.

Another feature of the way Project SEED is implemented and maintained might be attempted in community organizations in the process of replicating a program in many sites and as a way to insure that the intervention is true to the goals and strategies of the developers during evaluation. In Dallas the instructors were paid teachers but the system presumably operates the same in communities in which volunteers come into the classroom. On the one day each week Project SEED does not meet, SEED instructors have an in-service period to conduct discussions with the classroom teachers about the students and to observe and critique other SEED classes in progress. Conversely, SEED instructors and trainees on their in-service day are liable to visit other classes and provide a required critique of those teachers. Observation of this system during a previous evaluation in the Dallas school system led to a conclusion that the enthusiasm of Project SEED teachers was high and that this internal procedure maintained a very high quality and consistency of instruction (Webster & Chadbourn, p. 4).

Webster and Chadbourn report strong and consistent results for SEED, including an immediate impact from one semester of SEED on math achievement as measured by ITBS; a cumulative impact on achievement such that the more semesters of SEED instruction students had, the greater the differences in math achievement between SEED students and their matched comparisons; retention of math skills for four years such that students who participated in SEED for three years in grades four, five, and six still outperformed their matched comparisons on the Tests of Achievement and Proficiency in tenth grade; and former SEED students enrolled in significantly more higher math courses than their matched comparisons.

This evaluation offers several points for community organizations, even though most programs will not track students for four years after a three-year extensive intervention. First, even this major an intervention still had a cumulative impact, suggesting that if a program is excellent it would be difficult for it to reach "overkill." In the voluntary setting the outer limit probably is set by interest and willingness to participate. Second, this basically informal[11] approach had measurable and sustained impact on the existing standardized tests of math competency. That is, the impact of some experiences in informal science and math may need to await the development of tests and measures better suited to

---

[11]Although attendance at Project SEED sessions is required for all students in the designated classes, Project SEED is not treated as a school subject. In many communities, program leaders come from the community, not the school; students are full participants in the process of cooperative learning; and levels of participation, skills, and achievement are not graded.

capturing Resnick's "practical intelligence"—the ability to use what one has learned in everyday life—but evaluators of informal programs should not assume the absence of impact on existing tests. It may be possible to claim credit for change in achievement *if* these scores or grades can be obtained relatively unobtrusively. Third, the Project SEED evaluation is another in a continuing series in which positive outcomes seem to be linked with programs that correct for underexpectation. The success of fourth graders at high school math is labeled as such and interpreted to the participants as evidence of their abilities (Clewell et al., p. 204). It may be that success in meeting high expectations affects the attitudes and behavior of adults and participants alike.

EQUALS in Cleveland and Detecting Change

EQUALS was developed at the Lawrence Hall of Science to help classroom teachers sustain girls' interest in mathematics and to help them overcome gender stereotyping. Disseminated through teacher training, EQUALS now exists in many forms across the nation, in and out of schools. An Ohio EQUALS site was established at Cleveland State University; the resulting training and implementation of the program and concepts were evaluated by Rosemary E. Sutton and Elyse S. Fleming (1987; 1989). The authors argue for attempting systematic evaluation even when the conditions for experimental design are far from perfect. For example, they developed systems for teachers and principals to report on use of EQUALS materials and concepts as a less expensive alternative to classroom observation to document the implementation of the program. They analyzed teachers' attitudes before and after EQUALS training and found teachers more aware of gender discrimination and with an increased belief that they could affect the math classes girls take.

Much of the content of EQUALS focuses on problem-solving, and EQUALS teachers reported that they used more problem-solving activities in their classrooms. Sutton and Fleming (1987; 1989) used the Wisconsin Mathematical Problem Solving Test for students in grades 7–9 and created a similar test for students in grades 4–6. These are pencil-and-paper tests with a multiple-choice format—"story problems" that call on problem-solving strategies for which the student is unlikely to have learned a simple algorithm. In two years of evaluation, the mean scores of students in classes whose teachers had had EQUALS training increased significantly compared to the scores of students in non-EQUALS classes (1989). In one more episode in the Mathematics Attitudes Scales (MAS) story, Sutton and Fleming found that responses given by EQUALS students in grades 4–6 to an adapted Fennema-Sherman scale of math as a male domain became less stereotyped, while those of non-EQUALS students that age became more stereotyped. Also among these young students, though they tended to perceive less Utility of Math—as measured by a scale developed by Eccles (Parsons) et al. (1979)—over time, the EQUALS students declined less than the non-EQUALS students. On the remaining attitude scales used from the Fennema-Sherman and Eccles et al. instruments, no significant attitude change was detected for either age group (Sutton & Fleming, 1989).

Sutton and Fleming ponder the meaning of change in behavior (the problem-solving tests) that is greater than or precedes attitude change. They argue, "Attitude questionnaires and problem solving tests with high reliability and validity that can be administered easily to large groups do now exist. We urge other investigators to use them" (1989, p. 15).

Techniques of Assessment Revisited

While there is some remaining difference of opinion on the reliability and validity of the attitude scales *for assessing change in **attitudes** as a result of an intervention*, it certainly is the case that coordinated efforts to adapt, find, develop, use, and report on such scales are needed and helpful. Especially for programs offered by community organizations, whose budgets and resources for evaluation often are appropriately small, the standardization of basic attitude measures would be a boon.

The Sutton and Fleming contribution to assessing **performance** outcomes also is considerable. They used a test that is simple to administer and seems closely related to the goals and objectives of the program being evaluated. The test is a fairly long one, perhaps too long and school-like to be used in the out-of-school, voluntary context. Yet shorter versions might run the risk of precipitous losses of reliability. In any case, the search for measures of short-term gains in performance (desired behavior) that can be used on a before and after basis must surely continue. Still missing is a widely agreed-upon system for assessing the impact on science learning, again what Resnick calls **practical intelligence**, of programs in discovery science.

Girls Incorporated has considered developing pairs of activities that are different from each other but relate to the same scientific concept to be used on such a before and after basis (Girls Incorporated, 1992b). Taking apart a regurgitated owl pellet and later an abandoned bird's nest, for example, would allow participants to demonstrate their increased knowledge of habitat and ecological systems while demonstrating their increased sophistication in posing hypotheses and developing systems of classification. A system of observation then would be developed and tested to see if changes in the level of sophistication of the participants could be detected. A rating system might have:

- elements of **concept mastery** in the manner of the Assessment of Performance Unit (Gott et al., 1985), preferably with a scoring system that solves the problem of one right answer;
- elements of **familiarity** and **ingenuity** in using interesting stuff in the manner of Duckworth's measures of diversity and complexity (1978); and
- elements of **empowerment** and **equity** by noting styles of interaction among the participants and between the participants and the adult leader in the manner of the Girls Incorporated ethnographic study (Frederick & Nicholson, 1991).

Paired activities might also be used with a simplified version of Novak's **concept mapping,** based on interviews with participants immediately following their participation (Novak and Musonda, 1991; cf. Roth, 1992; Vargas & Alvarez, 1992). Also attractive is Spitzer and Foster's (1991) technique of **embedding** before and after (or, more precisely, earlier and later) assessments into the structure of the intervention by having students prepare a report for an external authority based on their recent accomplishments. The search for techniques to establish how (or if) informal science works must continue. Both the logic and the actual steps through which experience in informal science might expand the pipeline and encourage more young people to establish a longer and more intensive commitment to science are complex. Shared and agreed-upon measures of each step would lighten the burden and enlighten the program development of informal and formal science educators alike.

Next Steps for Community Programs

The search for informal science education programs to include in a chapter on community organizations yielded some outcome evaluations and many more programs in the early stages of or contemplating outcome evaluation. In informal science education, much remains to be learned about what works and why; knowledgeable people disagree about how to conduct evaluations and what constitutes evidence of a program's success. Despite the fact that there is no established pattern for conducting outcome evaluation of community-based programs, several of the program developers apologized that their programs were "still engaged" in process evaluation and had not yet "reached" outcome evaluation even after several years of operation. In general, community programs, especially those with outside funding, feel pressure to conduct outcome evaluation but reluctance to conduct evaluation that distorts the program's goals or swamps the participants.

Those who fear a precipitous leap into outcome evaluation probably have measured the risks well. But we do recommend orderly progress toward outcome evaluation so that a program is ready when the time and resources are right.[12]

Take Credit for the Evaluation You Do Already

In nearly every program, the developers and leaders have goals in mind, design activities that relate to the goals, observe the reactions of the participants and the leaders, and draw conclusions about whether the program is working as desired. The longer a program is in effect, the more likely the leaders are to have impressions of what the program is doing to or for the participants—how the participants are different after the program. They are conducting outcome evaluation, if only impressionistically. The shift to a systematic outcome evaluation is more a matter of recordkeeping than of mindset.

Write It Down, or, Better, Write It Up

Documentation is the first order of systematic evaluation. Program developers and implementers at every stage of program development can contribute to the general understanding of what works in informal science education and why it works by keeping careful track. Careful attendance records, participants' addresses for following them through school, session plans and materials lists to document the nature of the intervention, notes on changes in plans, and participants' reactions all are preliminary steps to asking whether participation in a program contributed to change in the knowledge, attitudes, and behavior of the participants. In addition, on the basis of these records, the program developers can answer many questions about whether the program is working, even if no skeptical outsiders are paying attention.

---

[12]Much of the following is based on the Girls Incorporated publication *Assess for Success* (1991). A similar orderly progression from documentation and goal-setting through attention to outcomes and eventually systematic outcome evaluation is outlined and recommended by Francine H. Jacobs (1988) in "The five-tiered approach to evaluation."

Writing it down serves the immediate needs of the people offering the program; **writing it up** can include an informal guide for the next program leader, a presentation for the community group sponsoring the program, or a paper for a professional conference or journal. The advantage of writing it up is that the insights of the program, including the travail of conducting evaluation, reach others who may benefit from the insights and avoid the pitfalls.

Evaluate No Outcome Before Its Time

People in the program business should always be evaluating, in the sense of reflecting on what they are doing and why. But systematic outcome evaluation consumes resources that might otherwise go into designing and conducting a program and so should be undertaken, we argue, only when some basic conditions are met:

- The goals of the program should be relatively clear.
- The desired outcomes should be clearly stated even if not readily measured.
- There should be a written record of what occurs in the program and an idea of how the events and activities are related to producing the outcomes.
- There needs to be a set of tools or strategies for documenting or measuring outcomes that seems to bear a strong relation to the desired outcomes (including short-term indicators of long-term outcomes).

At that stage, a decision can be made about whether the evidence of outcomes produced by a given evaluation design is worth the cost in time, money, good will, and other resources to conduct it.

Try Existing Tools to See if They Fit

Most of the community programs focus on one or more of the outcomes discussed here: persistence, performance, love of science, confidence, progress toward a scientific career, and similar concepts. A program seeking to document or measure outcomes can draw upon the experience of others with attitude scales, standardized tests and grade reports, protocols for observation, ideas for embedded assessment, and systems for tracking participants through time. Similarly, there is a range of models for establishing a comparison group as a way to increase the program's claim to having caused change in participants, from quasi-experimental to the opinion that such comparison groups are inappropriate and unnecessary. There is a great deal of room to try new approaches to evaluation and to invent new tools of assessment. But at least there is beginning to be a base from which to explore new paths.

Observe on the Buddy System

Some of the richest information about informal science education comes from qualitative studies. Observation of a program in action is difficult to replace as a component of documenting the nature of the intervention. Without detailed notes on the environment, the interaction of youth with adults, of youth with each other, of youth with the stuff and tools of science and technology, it is difficult to know what precisely is responsible for persistence or performance or confidence in science.

At a minimum, leaders can observe on the buddy system by visiting one another's programs and discussing the similarities and differences, strengths, and areas for improvement. Observation of the program with a set of guidelines on what to look for, with feedback to the leaders, is a system of evaluation but also seems to work well as a system of training new leaders to implement a program and of maintaining the quality of ongoing programs. Programs preparing to undertake outcome evaluation are fortunate if they are able to use observation to affirm the crucial characteristics of the program—content, process, interaction, types of participants, leadership styles, conceptualization of science. Audio- and videotaped records of observations may be especially helpful in reaching agreement among program developers, leaders, and evaluators about these crucial characteristics.

Measure Costs

A program that is a candidate for outcome evaluation almost certainly has been offered more than once, perhaps at more than one location or by more than one organization. Programs that track the basic costs of offering the program—the costs of staff time, training of staff, equipment and materials, space to conduct the program, and the like—can thus calculate average cost per participant. Even if cost is not the most important criterion of program developers, cost data can inform such decisions as the optimum size of program groups, the more economical of two similar strategies, and the resources needed to replicate a program elsewhere. That is, there are excellent reasons to measure costs during process evaluation. Programs lend themselves more or less well to relating costs to program outcomes. In the short run, it may seem risky to publish ostensibly hard cost data and soft outcome data about programs, lest uninformed funders or policy makers opt for the cheapest program over the most promising. As we become more proficient at stating and measuring outcomes, informal education may come to be recognized as a very cost-effective component of preparing the next generation of scientists.

Can We Talk?

At least until this volume, the literature on evaluating informal science was scattered and difficult to locate. Some of the community-based programs have had National Science Foundation support and thus an opportunity through meetings of project directors to discuss evaluation. Sessions on evaluation of informal science also have been held at meetings of the American Association for the Advancement of Science and the American Educational Research Association. These encounters form the basis for informal networks but much remains to be done. The field would move much faster if program developers, evaluators, and funders had more opportunity for oral and written communication. Readers interested in these issues should not wait for formal invitations to contact the programs and projects mentioned here.

Opportunities for Research and Evaluation

The advice in the previous section is meant to keep small and excellent programs of informal science, particularly but not exclusively community-based programs, engaged in the search for new understanding about what works but not pressured to prove their effectiveness. This section addresses the great potential in improving outcome evaluation in the larger enterprises and in cooperative ventures between evaluation researchers and the organizations delivering informal science education.

Developing New Tools

One theme throughout the chapter is that reliable and valid measures of inquiry-based science are still scarce. A few scales, such as the MAS, have been widely used, with success in some programs and perplexing results in others. Yet the search is still on for tools that assess the effectiveness of programs in conveying process-based "practical intelligence" in science and math. The best tools may not be simple and automatic to administer and so may be developed and used in separately funded research or in projects that bring together consortia of evaluators and program developers.

On the qualitative front, the commonality of desired outcomes among programs should facilitate the development, testing, and implementation of schemes for observation and documentation that transcend individual programs. We still do not know whether enthusiasm is more important than expertise, time with a given set of materials is more important than reflection on what one knows about them, or a mentor relationship is more important than a tutor relationship, in producing the outcomes of persistence and confidence. Much could be learned about expanding the scientific pipeline to include underrepresented groups from projects that work on documenting and evaluating programs with different strategies for achieving similar outcomes.

Which Attitudes Really Matter?

Implicit in the varied strategies of community-based organizations are assumptions about how to cause girls and young women, people of color, people with disabilities, and other targeted groups to remain in science. Several programs are based on the argument that to stay these groups must conceptualize science in a new way—as exploratory, imaginative, relevant, and fun. Others argue that the crucial attitude is confidence— this is something **I can do.** For still others the crucial task is to reduce stereotypes and include the participants in the groups who believe science is appropriate and socially acceptable for them to engage in. So far both the tools for precise measurement of these attitudes and connections between attitudes and behavioral outcomes, such as taking a degree in engineering, have been elusive. Yet most program implementers seem to believe that the attitudinal factors are important predictors of long-term persistence, performance, and progress toward a career in science. Research and evaluation can shed light on the relative importance of change in attitudes (compared to types and intensity of experience, changes in skills and behavior) to determining who stays in science.

Comparing and Contrasting Programs

Beyond the need for tools that are useful in many programs, we need comparative research that addresses the differences between programs in strategies and desired outcomes:

Persistence or performance? If programs focus on keeping people engaged in science (persistence) are they as likely to produce employed scientists as those that focus on good grades and test scores (performance)? Is persistence or performance a better outcome to measure for a program with a goal not of producing scientists but of creating a more scientifically literate populace?

Inquiry-based or curriculum-driven? To get students past high school and into college, is it more important for them to experience inquiry-based science or to be prepared for standardized tests? Should summer programs stress enthusiasm for doing science or preparation for the next semester's topics and skills in formal education? Or are both required?

Having a mentor or doing science? What are the critical periods to have a mentor? Is junior high school too late? Is a series of mentors as effective as a long-term relationship with one person? How important are the gender, racial and ethnic background, career status in science, or other characteristics of adults to their effectiveness as mentors and program leaders? What is the relative importance of a sustained relationship with adults and hands-on experience in science?

Early or late intervention? What are the critical times to intervene— very early, when children form concepts of science? By middle school, to assure the transition to algebra? In high school, to make college a real possibility? Early in college, to establish a pattern of high expectation for success? Is it ever too early or too late for a particular style or strategy of intervention? What are the signs that "the system" is changing so that more students will get encouragement in the formal educational system and be less reliant on informal education to sustain their interest and participation in science?

Measuring the Infrastructure

Community programs of informal science vary widely in the amount of time a participant spends in any one of them. The time some children in the early grades spend in informal programs might even surpass the time they spend on science in school. Other, much briefer, encounters with informal science—activities from a multimedia kit, visits to science centers and high technology companies—undoubtedly contribute to the "acculturation to science." How can the impact of the cumulative experiences in structured programs and occasional encounters be assessed? How much participation is enough? Until the formal educational system expands to make more young people feel they are scientific insiders, are informal structured programs **sufficient** for sustaining participation? **Necessary** but not sufficient? How do formal education, structured programs of informal science, visits to institutions of informal science, attention to science in the media, and relationships with parents and other adults fit together in sustaining young people's interest and participation in science? In short, what are the critical components of an effective infrastructure of informal science?

A resource base will be necessary to attract evaluation researchers to these larger questions and to entice community organizations with varied programs of informal science to participate in answering them. Fortunately, community organizations and others operating beyond the formal educational system are offering a broad spectrum of imaginative, thoughtfully developed, and faithfully implemented programs of informal science from which to learn.

Reflections on Informal Science in Community-Based Programs

For readers who are new to the community programs in science, the surprise may be the diversity and seriousness of the **programs,** not the assessments of their effectiveness. At least among those that have evaluated programs, community-based organizations have ambitious goals for expanding the scientific pipeline and the scientifically literate citizenry. Three models or pure types of community-based programs in science have been discussed. How might the society be different in the future if informal science programs of these three types were available to many more children and adults in the United States?

Discovery for Young and Old

In **discovery** programs people overcome stereotypes about science and scientists; they learn that they and everyone else can do science and that science is exciting, relevant and fun. So far the discovery programs are reaching only thousands, at most hundreds of thousands, of young people of school age. The adults in some of the discovery programs have little or no formal background in science; many of them must experience the surprise and joy of **doing** science and reflect on their own fears before they are able to facilitate the exuberant participation of young people. In the experience of community-based programs (e.g., Operation SMART, Linkages for the Future, Hands On Science Outreach, 4-H SERIES, FAMILY MATH and FAMILY SCIENCE) people of all ages are delighted to explore the natural world in the discovery mode. The key is helping people to think of science as interesting questions more than right answers, time to explore more than pressure to finish, interesting stuff more than intimidating equipment, all around us rather than dead and boring, and fun to share rather than a test of individual competence.

Ideally, of course, the discovery mode for learning science would be typical of formal education in science, so more people would have the confidence to stay in science in the first place. Until that day, discovery programs through community-based organizations could be expanded:

- to increase the positive experience in science of school-age young people, especially those currently underrepresented in science;
- to help marginally employed, unemployed and underemployed workers expand the types of education, training and work they are willing to consider;
- to assist parents in fostering their children's confidence and persistence in science;
- to increase the number of adults who have enough interest and confidence to pay attention to technically complex policy issues; or
- to encourage elementary teachers, including those who have little science background, to try more science in the classroom.

If even this partial list of benefits were realized, communities, foundations, corporations, and the National Science Foundation surely would be persuaded to offer further incentives to implement such programs. Soon every science museum, employer, religious institution, community center, and parks and recreation department would offer science discovery programs for several age groups. Then science would be the province of everyone, and the work force and the polity of 2020 would be better prepared.

Camp and Other Intensive Experiences in Science

Day camps and residential camps during the summer are designed to support the confidence and persistence of students who already have some interest in science. Especially the science camp has been offered to encourage underrepresented groups to remain in science at critical points of decision. Science camps for middle school girls encourage risk-taking and establish a peer group for whom being a whiz at computer software is worth the risk to one's social life if word gets out (EUREKA!). Math camp for Native American students (AISES Math Camp) similarly establishes a peer group and adult role models from several states and many tribes who think math problems are interesting and relevant to life. The science camps are carefully planned to be intellectually challenging, physically refreshing, confidence-building, and fun. Intensive experiences in controlled environments, fostering confidence and success in science, might profitably be offered in the following settings:

- for many more young people, especially those who need subsidized opportunities to participate, as summer alternatives to basketball, babysitting, or hanging around;
- in state parks, on cruise ships, and in elderhostels to make beginning and continuing interest in science accessible at every age and vacation schedule;
- for preschool and school-age children, as an optional benefit of their parents' employment, to increase the sense that science is an important use of discretionary time for children;
- for teachers, to focus on special science topics or targeted issues of equity in science;
- for high school students, as internships on college and university campuses, to increase confidence in science and the likelihood of staying in school;
- for students with disabilities, as part of special summer camps, Special Olympics, and life skills programs, as well as in the summer programs in which students with disabilities are mainstreamed;
- as a component of Job Corps, career readiness, and employers' staff development programs, to increase confidence and skills in problem solving; or
- for single parents, as subsidized day camps for the whole family, where everyone does science, uses computers, and gets a break from routine.

Of course some of these programs, from the NASA Space Camp to sophisticated university offerings, already exist. The task now is to make these opportunities accessible to many more people and to recognize that interest in science may need to be marketed to new groups. The old system of washing most people out of the pipeline will not be in the best interest of those of us who plan to retire in this century or early in the next.

Career Programs and Other Extensive Opportunities

The **career** programs are those most closely connected to the scientific pipeline. Most employ multiple strategies for identifying academically talented students and providing a system of challenges and supports to keep them moving through school and college, toward degrees and employment in science, mathematics, and engineering. In the career programs, young people from second grade on encounter older students and adults who support them through tutoring (Project Interface), rigorous study of science after school and on Saturday (Science Skills Center), interaction with role models who share their cultural heritage (Project Interface, Science Skills Center, MESA), preparation for high achievement on standardized tests (Science Skills Center), guidance in choosing and staying in college (MESA), and consistently high expectations. The career programs are based on the premise that a potential scientist or engineer may be entitled to as much support and encouragement as an athlete or musician. Certainly this is the case for talented students whose schools do not offer elementary science or high school physics because the resources are needed to secure the building. The minority engineering societies are exemplars of what can be accomplished when practicing scientists and engineers volunteer their time to prepare the next generation of professionals. Among the real-life situations the authors know about are:

- a veterinarian in Montana using road-killed animals to teach anatomy to a class of seventh graders;
- mathematicians from universities and industries using Socratic methods and hand signals to teach high school and college math to elementary students through Project SEED;
- women in engineering working with girls to build and race a soap box derby car;
- a university-based plant pathologist corresponding with a high school student about a term project;
- environmental scientists comparing data by computer with middle school students at a summer camp (TERC);
- African-American engineers pairing with high school students to be sure they take the Preliminary Scholastic Aptitude Test, sign up for calculus, and get a scholarship to a college or university with a strong engineering program;
- microbiologists working hard to recruit and retain a critical mass of women on the faculty and being sure undergraduates know this when they apply to graduate school;
- an Apple Computer Club donating software and members' time at a Girls Incorporated center; and
- a woman scientist who uses a wheelchair working with a science museum to develop an exhibit about accessibility.

For years the American Association for the Advancement of Science has been linking scientists through their professional societies with community-based organizations of all types. Yet many of the millions of young people still do not think of themselves as knowing one scientist well enough to call that person by name. The career programs remind us that, to paraphrase the African aphorism, it takes a community of science to rear a scientist.

Last Word

The community-based programs in informal science education will continue to ask the challenging questions about how to encourage young people to pursue science. They will continue to assess outcomes in increasingly sophisticated ways. Meanwhile, the results of evaluations of discovery programs, science camps, and career programs indicate that these are promising models both for expanding the scientific pipeline and increasing the number of citizens who are interested in and pay attention to science.

Summary

Many community-based programs have discovered how to involve young people, especially those currently underrepresented in science, meaningfully and joyfully in the challenges of scientific and mathematical thinking. In this, they are far ahead of many school systems in fostering cooperative learning and hands-on experience leading to critical thinking and an ability to solve problems. Community-based programs in science, mathematics and technology have attempted to improve—and measure—such outcomes as experience, equity, persistence, performance, love of science, confidence in scientific ability, and progress toward a career in science.

The number of outcome studies of informal science education in community-based programs in out-of-school contexts is fairly small. Programs that have been evaluated can be divided by structure and purpose into three groups:

Discovery programs—after-school programs with occasional participation in informal science and a focus on making engagement in science fun and rewarding. These programs are designed to make science fascinating and exhilarating by having young people and sometimes their parents engage directly with the stuff of science and then reflect on their experiences. Discovery programs are offered by youth organizations such as Girls Incorporated (Operation SMART), Girl Scouts (Linkages for the Future) and 4-H (SERIES). The programs

FAMILY MATH and FAMILY SCIENCE are offered by community organizations such as the National Urban Coalition and the Alba Society, and Hands On Science Outreach becomes its own community organization as new communities adopt it.

Implicit or explicit in many of the discovery programs is that these informal programs are competing with school for the opportunity to define science in order to retain enthusiasm for engaging in it. The outcome studies thus have focused on positive attitudes toward science and intentions to persist in science in school and beyond. Attitude measures have included the Draw-A-Scientist Test to see if participation in the program leads to less stereotyped views of scientists, and the Math Attitudes Scales to see if attitudes measured by such subscales as anxiety, usefulness, and parental support have changed.

Science camps—intensive short-term programs, often residential or day camps, with course-like immersion in science and math and a focus on providing participants with the confidence and competence to pursue formal education in science. Often these programs recruit participants who are capable of good work in science, pre-engineering or computers but who are at risk of falling out of the scientific pipeline. Most are taught by people who earn their living in formal education and who share gender, race or culture with the participants. A goal is to provide an engaging experience in science out of school that will lead to persistence in science in school, persuading the participants that they can legitimately see themselves as scientific insiders rather than outsiders. Among the evaluated science camps are the American Indian Science and Engineering Society (AISES) Summer Math Program; EUREKA!, a math, science and sports camp developed by the Women's Center at Brooklyn College; Mount Holyoke SummerMath; the University of Michigan Summerscience; and TERC Environmental Network Project.

In addition to embedded measures of science learning and before-and-after measures of attitudes toward science, the evaluations of science camps have focused on the participants' intentions to pursue science and math in school and their actual enrollment and performance in these subjects.

Career programs—longer-term programs with frequent participation and multiple strategies, including targeted contact with science professionals and a focus on preparing participants for careers in science. The career programs tend to be multifaceted support systems designed to insure that the participants remain in the path toward a designated job or career. The pure type is extensive, with participation at least once a week for a year or more, often with a presumption of continuing involvement through college and beyond. Among the evaluated career programs are Project Interface in Oakland, CA, MESA across California, the Science Skills Center in Brooklyn and, in several communities, Project SEED.

Evaluations of the career programs generally compare the science and math performance of participants with the performance of their classmates who have not participated. Several of the programs have elaborate systems for tracking the educational and career paths of their participants over time to assess the program's success in retaining participants in the scientific pipeline well beyond the duration of the intervention.

The research designs range from simple and basic to ingenious and sophisticated, from qualitative to quantitative. Developers of community-based programs can increase the amount and sophistication of evaluation they do in stages, beginning with a clear statement of their purposes and the outcomes they hope their programs are producing for the youth or adults who participate. The community-based programs will continue to ask the challenging questions about how to encourage young people to pursue science. They will continue to assess outcomes in increasingly sophisticated ways. Meanwhile, the results of evaluations of discovery programs, science camps, and career programs indicate that these are promising models both for expanding the scientific pipeline and increasing the number of citizens who are interested in and pay attention to science.

Acknowledgements

This chapter is in every way a collegial effort. We appreciate the opportunity to write about community-based programs afforded by Valerie Crane and Tom Birk of Research Communications Ltd. and Hyman Field of the National Science Foundation. Their clarity, support, flexibility, and enthusiasm made the process enlightening and enjoyable. Dozens of colleagues made special efforts to share their knowledge and resources. Among the most generous were Ralph Buice, Beatriz Chu Clewell, Cinda Sue Davis, Catherine Didion, June Foster, George Hein, John Hoover, Laura Jeffers, Phyllis Katz, Nancy Kreinberg, Carole Lacampagne, Laura Martin, Marsha Lakes Matyas, Charles Miller, Jeff Miller, Pedro Pedraza, Richard Ponzio, Steven Rogg, Mark St. John, Harris Shettel, Belinda Seto, Rosemary Sutton, and Bonnie Van Dorn. Indeed this spirit of cooperation and generosity is characteristic of the expanding network of informal science education.

It is in the context of the Girls Incorporated initiative Operation SMART that much of our thinking about evaluation of informal science has been developed and colleagues Ellen

Wahl, Libby Palmer, Alice Miller, Melanie Flatt, and previously Julie Frederick, Julie Hamm, Rachel Theilheimer, Jane Quinn, Jenni Martin, and many others have contributed to our understanding.

Faedra Lazar Weiss and I worked together for the entire process from outline through completion. Although she drafted the abstracts and I drafted the body of the chapter, the result is joint authorship in its fullest sense. Patricia B. Campbell drafted the section on EUREKA! and is a full partner in our thinking and conceptualization about evaluation in informal science education. Mary Maschino, librarian/media specialist at the Girls Incorporated National Resource Center, performed much of the work of locating and acquiring elusive materials.

The views expressed do not necessarily reflect positions of Research Communications Ltd. or the National Science Foundation and any errors of fact or interpretation are my responsibility.

Heather Johnston Nicholson

References

American Association for the Advancement of Science [AAAS]. (1989). *Project 2061: Science for all Americans*. Washington, DC: Author.

American Association for the Advancement of Science. (1992). *Girl Scouts and science: In touch with technology* [interim report]. Washington, DC: Author, The Directorate for Education and Human Resource Programs.

California Postsecondary Education Commission [CPEC]. (1992). *Final report on the effectiveness of intersegmental student preparation programs*. Sacramento, CA: Author.

Campbell, P. B., & Shackford, C. (1990). *EUREKA! participant follow up analysis*. (Available from Patricia B. Campbell, Ph.D., Campbell-Kibler Associates, Groton Ridge Heights, Groton, MA 01450).

Campbell, P. B., & Shackford, C. (1992). *EUREKA!; 1992 evaluation report*. (Available from Patricia B. Campbell, Ph.D., Campbell-Kibler Associates, Groton Ridge Heights, Groton, MA 01450).

Carnegie Corporation of New York. (1992). *A matter of time: Risk and opportunity in the nonschool hours* [Executive summary]. New York: Author.

Carpenter, E. L. (1992, November 30). National Chemistry Week: An off-year success for ACS. *Chemical & Engineering News*, pp. 29–30, 32–33, 36–40, 42, 44, 46.

Center for the Study of Testing, Evaluation, and Educational Policy, Boston College [CSTEEP]. (1992). *The influence of testing on teaching math and science in grades 4–12*. Chestnut Hill, MA: Author.

Chambers, D. W. (1983). Stereotypic images of the scientist: The draw-a-scientist test. *Science Education, 67*, 255–265.

Clewell, B. C. (1992). *Final report: First year evaluation of the Eureka! teen achievement program*. Report submitted to EUREKA!, Women's Center of Brooklyn College, Brooklyn, NY.

Clewell, B. C., & Anderson, B. (1991). *Women of color in mathematics, science & engineering: A review of the literature*. Washington, DC: Center for Women Policy Studies.

Clewell, B. C., Anderson, B. T., & Thorpe, M. E. (1992). *Breaking the barriers: Helping female and minority students succeed in mathematics and science*. San Francisco: Jossey-Bass.

Duckworth, E. (1978). *The African Primary Science Program: An evaluation and extended thoughts.* Grand Forks, ND: University of North Dakota, North Dakota Study Group on Evaluation.

Eccles, J. et al. (1979). *See* Parsons.

EQUALS. (1992). *FAMILY MATH and Matemática para la familia.* Berkeley, CA: Lawrence Hall of Science, University of California at Berkeley.

Etaugh, C., & Liss, M. B. (1992). Home, school and playroom: Training grounds for adult gender roles. *Sex Roles, 26,* 129–147.

Federal Coordinating Council for Science, Engineering, and Technology: Committee on Education and Human Resources, Public Understanding of Science Subgroup [FCCSET-CEHR PUNS]. (1992). *Report on the expert forum on public understanding of science* [draft]. Forum conducted in Alexandria, VA, August 20–21, 1992.

Fennema, E., & Sherman, J. A. (1976). Fennema-Sherman Mathematics Attitudes Scales: Instruments designed to measure attitudes towards the learning of mathematics by females and males (MS.1225). *JSAS Catalog of Selected Documents in Psychology, 6*(2), 31.

4-H SERIES Project. (1992). *4-H SERIES Project: NSF directors' meeting* [project folder]. (Available from Richard C. Ponzio, Ph.D., Director, 4-H SERIES Project, University of California Cooperative Extension, 4-H Center, Davis, CA 95616-8599).

Frederick, J. D., & Nicholson, H. J. (1991). *The explorer's pass: A report on case studies of girls and math, science and technology.* Indianapolis: Girls Incorporated.

Fry, C. (1990, April). *Sex-related differences in mathematical achievement: Learning style factors.* Paper presented at the annual meeting of the American Educational Research Association, Boston, MA.

George, M. (1990, 27 June). Camp helps girls imagine new careers. *Detroit Free Press,* 1B, 3B.

Gibbons, A. (1992). Future conditional I: Minority programs that get high marks. *Science, 258,* 1190–1191, 1194–1196.

Gilliom, M. E., Helgeson, S. L., & Zuga, K. F. (1993). Science, technology: Opportunities [special issue]. *Theory Into Practice, 31*(1).

Girls Clubs of America [Girls Incorporated]. (1988). *Operation SMART® research tool kit: A package of program evaluation activities for girls.* Indianapolis: Author.

Girls Incorporated. (1990). *Spinnerets and know-how: Operation SMART® planning guide.* New York: Author.

Girls Incorporated. (1991). *Assess for success: Needs assessment and evaluation guide.* New York: Author.

Girls Incorporated. (1992a). *Operation SMART® research tool kit: A package of program evaluation activities for girls. Supplement to the staff handbook.* Indianapolis: Author.

Girls Incorporated. (1992b). *Research means look again: Concept paper for tools to assess informal science.* Indianapolis: Author.

Gong, B., Lahart, C., & Courtney, R. (1991). *Current state science assessments: Is nothing better than something?* [Research Report RR-91-2]. Princeton, NJ: Educational Testing Service.

Goodman, I. F. (1992). An evaluation of the Hands On Science Outreach Program. In *How are we doing? A progress report of an evaluation of the Hands On Science Outreach, Inc. (HOSO) program.* Rockville, MD: Hands On Science Outreach.

Goodman, I.F. with Rylander, K. (1992). *An evaluation of children's participation in the Hands On Science Outreach program.* Rockville, MD: Hands On Science Outreach.

Gott, R., Davey, A., Gamble, R., Head, J., Khaligh, N., Murphy, P., Orgee, T., Schofield, B., &

Welford, G. (1985). *Science in schools ages 13 and 15: Report No. 3.* London: Assessment of Performance Unit, Department of Education and Science.

Hardin, J., & Dede, C. J. (1992/1973). Discrimination against women in science education: Even Frankenstein's monster was male. In M. Wilson (Ed.), *Options for girls: A door to the future* (pp. 232–240). Austin, TX: Pro-Ed. Reprinted from *The Science Teacher, 40*(9), 18–21.

Harty, H., Kloosterman, P., & Matkin, J. (1989). Science hands-on teaching-learning activities of elementary school teachers. *School Science and Mathematics, 89,* 456–467.

Hazen, R. M., & Trefil, J. (1991). *Science matters: Achieving scientific literacy.* New York: Doubleday.

Heath, S. B. (1983). *Ways with words: Language, life, and work in communities and classrooms.* Cambridge: Cambridge University Press.

Hoover, J. J., & Abeita, C. (1992). *Effects of an intensive summer program on attitude toward mathematics of American Indian eighth grade students.* Unpublished report. (Available from John J. Hoover, Ph.D., American Indian Science & Engineering Society, 1630 30th Street, Suite 301, Boulder, CO 80301-1014).

Johnson, M. (1991). Assessing accelerated science for African-American and Hispanic students in elementary and junior high school. In G. Kulm & S. M. Malcom (Eds.), *Science assessment in the service of reform* (AAAS Publication 91-33S) (pp. 267–282). Washington, DC: American Association for the Advancement of Science.

Kahle, J. B. (1987). Images of science: The physicist and the cowboy. In B. J. Fraser & G. J. Giddings (Eds.), *Gender issues in science education* (pp. 1–11). Perth, Western Australia: Curtin University of Technology.

Kahle, J. B. (1989). Images of scientists: Gender issues in science classrooms. In Key Centre for School Science and Mathematics, *What research says to the science and mathematics teacher: Number 4* (pp. 1–7). Perth, Western Australia: Key Centre, Curtin University of Technology.

Katz, P. (1992). Preface. In *How are we doing? A progress report of an evaluation of the Hands On Science Outreach, Inc. (HOSO) program.* Rockville, MD: Hands On Science Outreach.

Kelly, A. (1991). *Women and mathematics report: July 1, 1990 through June 30, 1991.* (Available from Author, Department of Mathematics, Santa Clara University, Santa Clara, CA 95053).

LaFollette, M. (1992/1988). Eyes on the stars: images of women scientists in popular magazines. In M. Wilson (Ed.), *Options for girls: A door to the future* (pp. 40–54). Austin, TX: Pro-Ed. Reprinted from *Science, Technology, & Human Values, 13*(3–4), 262–275.

Lapointe, A. E., Mead, N. A., & Phillips, G. W. (1989). *A world of differences: An international assessment of mathematics and science* [Report No. 19-CAEP-01]. Princeton, NJ: Educational Testing Service.

Linkages for the Future . . . it's working! (1992, Spring). *Girls and Science: Linkages for the Future,* p. 1.

Linn, M. C., & Hyde, J. S. (1989). Gender, mathematics and science. *Educational Researcher, 18*(8), 17–19, 22–27.

Martin, L., Ascher, C., & St. John, M. (1992). *Is there science and math after school?* Paper presented at the American Educational Research Association meetings, San Francisco, April 1992.

Mason, C. L., Kahle, J. B., & Gardner, A. L. (1991). Draw-a-Scientist Test: Future implications. *School Science and Mathematics, 91,* 193–198.

Massey, W. E. (1992). Past imperfect I: A success story amid decades of disappointment. *Science, 258,* 1177–1179.

Miller, A. (1990). *EUREKA teen achievement program—July 23–August 21, 1990: Final evaluation.* (Available from Author, Women's Center, Brooklyn College, Chestnut Hill, MA).

Mount Holyoke College. (1991). *SummerMath* [program brochure]. South Hadley, MA: Author.

Murphy, P. (1989). Gender and assessment in science. In P. Murphy & B. Moon (Eds.), *Developments in learning and assessment* (pp. 323–336). London: Hodder & Stoughton.

Murphy, P., & Gott, R. (1984). *Science assessment framework age 13 & 15* [Science report for teachers: 2]. Hatfield, Hertfordshire, England: Association for Science Education.

National Collaboration for youth. (1990). *Report on the nationwide project Making the Grade: A report card on American youth.* Washington, DC: Author.

Nelson, K. (1990, 27 June). Girls get head-start in science. *Detroit News* (Western Wayne), 3B.

Nicholson, H. J. (1992). *Gender issues in youth development programs.* Indianapolis: Girls Incorporated.

Nicholson, H. J., & Hamm, J. K. (1990). *Measuring equity: In search of tools with sandwich validity.* Indianapolis: Girls Incorporated.

Nicholson, H. J., & Matyas, M. L. (1988). *Evaluation report: MuseumLink Project.* Indianapolis: Girls Clubs of America [Girls Incorporated].

Novak, J. D., & Musonda, D. (1991). A twelve-year longitudinal study of science concept learning. *American Educational Research Journal, 28,* 117–153.

Oakes, J., Ormseth, T., Bell, R., & Camp, P. (1990). *Multiplying inequalities: The effects of race, social class and tracking on opportunities to learn mathematics and science.* Santa Monica, CA: RAND.

Pagni, D., & Long, V. M. (1993). Targeting girls: MISS. *The Mathematics Teacher, 86*(1), 95–96.

Parsons, J., Adler, T., Futterman, R., Goff, S., Kaczala, C., Meece, J., & Midgley, C. (1979). *Self-perceptions, task perceptions, and academic choice: Origins and change* [Final report to the National Institute of Education, Washington, DC]. (ERIC Document Reproduction Service No. ED 186 477).

Pechman, E. M. (1988). *Building the case for developmentally responsive mathematics programs for young adolescents.* Carrboro, NC: Center for Early Adolescence.

Project Interface. (1987a). *Class placement of students for 1986–87 and 1987–88* [unpublished data]. (Available from Author, J. Alfred Smith Fellowship Hall, 8500 "A" Street, Oakland, CA 94621).

Project Interface. (1987b). *1982–83 through 1986–87 results on the Comprehensive Test of Basic Skills (CTBS test)* [unpublished data]. (Available from Author, J. Alfred Smith Fellowship Hall, 8500 "A" Street, Oakland, CA 94621).

Project Interface. (1987c). *Summary report on 9th grade graduates* [unpublished data]. (Available from Author, J. Alfred Smith Fellowship Hall, 8500 "A" Street, Oakland, CA 94621).

Resnick, L. B. (1987). *Education and learning to think.* Washington, DC: National Academy Press.

Rogg, S. R. J. (1991). *Evaluation of year one implementation of Operation SMART in Rural Communities by the Girls Club of Rapid City, Inc.* Report submitted to the Bush Foundation.

Roth, W.-M. (1992). Dynamic evaluation. *Science Scope, 15*(6), 37–40.

St. John, M., & Shields, P. M. (1988). An expert mini-conference: Exploring the assessment of learning in informal science settings. In M. S. Knapp, P. M. Shields, M. St. John, A. A. Zucker, & M. S. Stearns, *An approach to assessing initiatives in science education* [SRI Project No. 1809] (pp. 83–103). Menlo Park, CA: SRI International.

Shettel, H. (1991). Summary report: An evaluation of the Hands On Science Outreach Program. In *How are we doing? A progress report of an evaluation of the Hands On Science Outreach, Inc. (HOSO) program.* Rockville, MD: Hands On Science Outreach.

Spitzer, W., & Foster, J. (1991). *The TERC Environmental Network Project pilot summer program: Final evaluation report.* Cambridge, MA: TERC.

Sutton, R. E., & Fleming, E. S. (1987). *EQUALS at Cleveland State University: 1985–6 evaluation report.* (Available from Rosemary E. Sutton, Cleveland State University, Euclid Avenue at East 24th Street, Cleveland, OH 44115).

Sutton, R. E., & Fleming, E. S. (1989). Evaluating an intervention program: Results from two years with EQUALS. (Available from Rosemary E. Sutton, Cleveland State University, Euclid Avenue at East 24th Street, Cleveland, OH 44115).

Trueba, H. T., & Delgado-Gaitan, C. (1985). Socialization of Mexican children for cooperation and competition: Sharing and copying. *Journal of Educational Equity and Leadership, 5*(3), 189–204.

Vargas, E. M., & Alvarez, H. J. (1992). Mapping out students' abilities. *Science Scope, 15*(6), 41–43.

Webster, W. J., & Chadbourn, R. A. (1992). *The evaluation of Project SEED: 1990–91* (EPS91-043-2). Dallas: Dallas Independent School District, Department of Evaluation and Planning Services.

Weisbaum, K. S. (1990). *"Families in FAMILY MATH" research project (NSF proposal no. MDR-8751375): Final report.* Berkeley, CA: Lawrence Hall of Science, University of California, Berkeley.

Zuckerman, H. (1987). Persistence and change in the careers of men and women scientists and engineers: A review of current research. In L. S. Dix (Ed.), *Women: Their underrepresentation and career differentials in science and engineering* (pp. 123–156). Washington, DC: National Academy Press.

5

CHAPTER

Understanding the Dynamics of Informal Learning

Valerie Crane, Ph.D.

What the Research Says

As previous chapters suggest, *achieving impact* for any informal learning enterprise is dependent on several project conditions:

- A reasonable set of assumptions about what the project could accomplish.
- A project design that will produce intended effects.
- A production effort that implements the design and matches project goals.
- An adequate understanding of the target audiences.

When these project conditions are not met, the likelihood of significant impact is seriously threatened. When impact studies show no effects, it is important to understand why. Without a deeper level of understanding of the underlying causes of success and failure, the evaluation effort itself will fail.

It also is important that the following research conditions are met for *assessing impact*.

- Having a design that is adequate to the task of identifying impacts.
- Developing measures that match project goals and are sensitive to impacts.
- Executing the study according to acceptable standards of research.

This volume reviewed studies which suggest that informal science learning experiences can make a difference among the populations they reach. Others failed to show intended or unintended effects. In some cases there were no effects because the above mentioned project conditions were not met, and in other cases, the required research conditions were not met. The research challenges varied depending on the informal learning setting.

The fewest number of *impact* studies have been conducted in the area of television; the environment that may well be the most difficult of the three media to study and measure. It also is the most expensive to produce, involving substantial investment of public resources, although that expense is offset by high efficiency (i.e., reaching more people at a lower cost than either museums or community-based projects). It is both notable and disappointing that the medium that has been heralded by many to have had the greatest impact on our lives as a communications vehicle in the past four decades has been studied the least. Chen appropriately calls for more research on this important medium.

The museum field has produced a more substantial number of research studies than has either television or community-based projects. The higher level of research activity compared to other informal learning settings may be a function of the proximity of the exhibit designer to the public. It is difficult for exhibit designers to ignore public enthusiasm or disdain for exhibits they have created when they see these responses where they work. The museum experience also provides a rich opportunity to observe how the informal learning process works in a relatively unstructured context. Meaningful insights can be gained on how the public interacts with exhibits simply by walking around and observing the interactions of visitors. Similar observation of television viewing would be more obtrusive, making it difficult to capture the "real" learning environment. However, museums have conducted far fewer summative or impact studies than formative and developmental studies.

Community-based projects provide yet a different environment that is voluntary by nature but more structured than either the television or museum environment. Community-based projects are rich laboratories for research as well, since more variables can be controlled and the learning process can be observed in an environment over a longer period of time.

It is interesting that little to no research exists on what the additive effects of these three media are on informal science learning, although the public does not experience these events in isolation, as illustrated by the case of Cindy in our introductory chapter. The need for research which looks at the cumulative impacts of informal learning will be discussed later in this chapter as we explore the agenda for continued research and evaluation in the field of informal learning.

While each of the chapters provides at least some evidence that informal learning can make a difference, one is also struck by:

1. *how difficult those impacts or effects are to measure;*
2. *the extent to which impacts or effects are often not known or well understood; and*
3. *the lack of systematic study of how the public experiences informal learning.*

In order to examine these and other issues that professionals face in informal science learning research, the author conducted two dozen interviews of one to three hours in length with informal learning researchers in the television, museum, and community-based project fields and informal science with practitioners (science museum directors, museum exhibit developers, television producers, community-based project developers). A list of interviewees is included in the Foreword of this volume. The interviews were open-ended and focused on what these professionals believe to be the future directions that informal science learning research should take.

The interviews reflected similar concerns among respondents despite both the relatively unstructured format for the interview and the diverse backgrounds of those interviewed. This was notable since, in the real world, there are few opportunities for these professionals to interact, especially across media (a common complaint among all of them).

This chapter is devoted to synthesizing and integrating the ideas shared by these individuals. Any opinions, findings, and conclusions or recommendations expressed in this material are those of the authors and the professionals who participated in the interviews, do not necessarily reflect the views of the National Science Foundation, and are organized in such a way to reflect where they think we are and where we need to go.

Goals for Informal Learning

Setting Goals

Informal science learning was operationally defined in Chapter One as referring to activities that occur outside the school setting, are not developed primarily for school use, are not developed to be part of an ongoing school curriculum, and are characterized by voluntary as opposed to mandatory participation as part of a credited school experience.

With this definition of informal learning in mind, those interviewed usually began their comments with a clear call for defining what the goals of informal learning should be.

A common theme in the discussion about goals was the need to come to an agreement on what science is. While some defined science as a set of disciplines and a body of content that needed to be mastered, many more felt that science was a way of looking at the world and that informal science learning should focus on providing the audience with ways of getting to the answers rather than on the answers themselves.

One researcher called for something similar to the National Council on the Teaching of Mathematics (NCTM) standards that would focus informal science and math learning on teaching problem-solving rather than on computation and getting the right answers. This approach would move the research field away from the multiple-choice tests used by Miller (1987) to measure "science literacy" and toward measures of personal experience, science concepts, self-confidence, facts, and problem-solving skills. The move away from a construct of science literacy where there is a minimal required body of knowledge is especially appropriate for informal learning since informal learning is not curriculum-driven, cannot expect uniform exposure to the stimulus, and, therefore, cannot expect uniform outcomes. As the field of science continues to grow, it becomes increasingly difficult to define the body of content we all "should" know.

While talking about the goals for informal science learning, a number of respondents discussed the mission of the informal science learning institutions themselves. On one level, the mission of the institutions is to make science directly available to the public. Mark St. John (1993), for example, refers to this accessibility of resources as fulfilling an important societal need and suggests that this should be the focus of evaluation as much as any project-specific outcomes. Because of this, he suggests that we "evaluate" science museums in the same way as libraries. He sees these institutions as part of an infrastructure that supports public understanding of science and, thus, evaluation criteria should include design characteristics (e.g., utility, accessibility, compatibility with other resources, flexibility), use (e.g., level, audiences, unique functions), perceived value (e.g., quality, value, benefits, fulfilling market demands, and user needs), and societal benefits (e.g., cost/benefit ratio, range of functions, etc.). This concept of infrastructure and support systems strongly suggests that the distribution, promotion, and packaging of informal science learning programs should be evaluated for their appropriateness and effectiveness in reaching the target audiences, as noted in the introductory chapter of this volume.

Respondents also discussed the role of the funder in the research process. The motivations for funding research in the informal science learning domain will set the agenda for what is to come. Who asks what questions, under what conditions, and for what purposes? Studies conducted to justify funding decisions or to assess the effectiveness of a project (evaluation) look very different from studies designed to discover how informal science learning experiences work to serve the public (research).

Another debate that directly impacts the nature and direction of research focuses on who the target audiences of informal science learning should be. For example, Miller (1987) claims that the level of science literacy, as he defines it, among the American public is very low (5–7%). Miller proposes that public resources (including those that go to museums and public television) should be focused on the 20% of the population that already participates to some degree in informal science experiences. With this emphasis, exhibitry and programming would target these audiences rather than the general public.

In fact, museums and public television have, at least in the past, attracted a more upscale audience than characterizes the general population. But many informal science practitioners object to this goal for informal science education. They feel that accessibility to a broader public is an essential part of their mission. If this is the case, then the match between offering and audience should be carefully examined. This returns us to a question that was raised in the first chapter. Are we designing and distributing informal science learning experiences that can and will reach the general public? If we are not having an impact on the level of science literacy among the general public, is it because the public cannot be educated or because the informal learning experiences target a college-educated, science-literate audience?

Learning about Learning

There were two areas of investigation that respondents felt were important for the advancement of research in informal science learning. Almost all respondents felt that there was a need for ongoing research efforts which would provide a theoretical framework for understanding the learning process. In the Preface of this book, Tressel calls for a better understanding of how people use and integrate information from informal learning sources. The second area of research would be to achieve a better understanding of the perceived value and impacts of specific

informal learning projects as a way of advancing knowledge in the field
of informal science learning.

The constructivist approach to informal learning discussed by Chen
in Chapter Two is a particularly appropriate one for informal learning
because of the interconnections that one makes as one moves through a
variety of learning experiences, as did our third-grader Cindy in Chapter
One. How does Cindy integrate the information she acquires over time?
How does she make meaning out of the different inputs? In this context,
learning from formal and informal learning sources is blurred. It is
possible that schools systematically have been claiming gains in learning
which result from much more than the inputs in the classroom over the
course of a year; in fact, the cumulative impacts are most obviously a
combination of what children learn in and out of school.

Studies that provide the conceptual or theoretical framework for
informal learning are more research driven than evaluation driven. This
type of research is based on hypotheses and is more likely to lead to a
conceptual framework for understanding informal learning. Without
such a framework, evaluation will continue to focus on the pragmatics of
audience composition, scale, and communication rather than on the
deeper indicators of impact.

During interviews regarding gaining knowledge about informal science learning, these concerns were expressed in different ways:

- *"We need to learn how people learn, and what people do when they
  are learning, not just what they learn."*
- *"Science is about process—analyzing, synthesizing, applying. How
  do we measure that?"*
- *"Informal learning is a conduit to other layers of learning. How can
  we demonstrate that?"*
- *"How can we build a system in which informal and formal education
  can thrive?"*
- *"How can informal and formal learning complement each other and
  lead to an enthusiasm for lifelong science learning?"*

There was consistent support for research efforts which address these
questions and concerns.

Project-Based Research

Advocates for project-based studies were focused on practical, real-
world considerations. There is a strong sense that as projects are evaluated, the findings will begin to show how informal science learning
projects work, under what conditions, and for what audiences. These

kinds of questions are essential for planning, designing, and executing successful (and continuously more successful) informal learning experiences.

There also was criticism of the project-based approach, however. There was a concern that too much research at the present time gets "hung up" on the stimulus rather than the learner and the learning process. A caveat for project-based research, then, is to expand the vision beyond the stimulus and to gain a larger sense of the learning experience.

When discussing project-based research, many commented that we need to develop a common sense of what informal learning projects can and cannot do. A framework is needed that is clear to developers, writers, the audience, and educators. In a series of studies by Crane (1992), producers consistently responded to television news stories on science in a significantly different way than did the target audience. These studies showed that producers and viewers respond to news in the same way only about 50% of the time. This means that the capacity of the speakers (producers) to see information and respond in the same way as the viewers is at chance level. When one considers the importance of the relationship between the producers and audience, this statistic is, indeed, startling. Informal learning practitioners should consider ways of improving this relationship between producers and their target audiences in order to communicate more effectively. A better knowledge of how people learn from informal science learning sources would be important in this process.

Another point of discussion among many of those interviewed was the need for projects to define what they would consider the "success" of the project. Respondents expressed the need to specify objectives which:

- are clear, unambiguous, conceptually consistent, and realistic;
- are not shallow or trivial because they are easier to accomplish or measure; and
- fit into a cohesive framework for learning.

A tension exists between this call for realism and avoidance of the trivial. The deeper meanings of a science experience are difficult to define, much less to measure. Respondents felt it was more appropriate and realistic to get kids excited about science than to impart a body of science facts that would be the subject of a content test. But then how do you measure those more important but subtle outcomes?

One dimension of setting realistic and meaningful goals for a project requires a firm understanding of the medium that conveys the message.

What can each of these media accomplish that others cannot? What are the advantages and limitations of each?

Television is good at illustrating process and phenomena, but is typically superficial and less effective at introducing entirely new constructs of knowledge. One respondent put it well, "Television is at its best when children use math and science as verbs and not nouns." Museums, on the other hand, are good at creating concrete examples and providing an interactive experience but less effective at dealing with large-scale issues which are artifact poor. Community-based projects can provide highly personalized experiences over extended periods of time, target special audiences, and provide face-to-face teaching, but are correspondingly more labor intensive and costly than a medium like television, for example.

Research Challenges in Informal Learning

At the outset of the book we discussed why informal science learning is an important part of the overall educational landscape. Similarly we need to establish why research is an important part of the landscape and how it contributes to the enterprise.

The professionals interviewed for this chapter felt that it is not enough for informal science learning projects to exist. A number wanted to know whether informal science learning institutions are carrying out their mission in the most efficient manner. Research is needed to determine whether projects contribute to this mission.

An important role of the research endeavor is to determine whether projects work in the ways they were intended to work and also to determine whether there are important unexpected outcomes that might not have been planned.

One example of unintended outcomes came in an evaluation of a television program about legal rights as they relate to child sex abuse cases in a foreign country where there is almost no public discourse on this issue. The original intent of the program was to inform the audience about their legal rights. But an unexpected outcome that intrigued the production team more than the acquired knowledge of their rights, was the finding that the audience felt empowered to discuss the issue for the first time among their peers, and that they had learned from the program how to have a meaningful dialogue on this subject. So an unexpected outcome was the increase in frequency of public discussion of the issue after watching the program. Research designs need to be sensitive to these outcomes as well as to those that are planned or expected.

In the following sections, important issues that influence what research is undertaken, the measurement challenges that exist in informal learning research, and the linkages that need to be made across informal experiences and with the schools are discussed.

Toward a Theory of Informal Learning

As stated earlier, a consistent question from those interviewed was:

What are the theories and principles that guide informal learning?

Whether practitioners, researchers, or policy makers, respondents felt that informal learning was, as one put it, "undertheorized." Without conceptual building blocks it is difficult to plan and assess projects. Many felt that research could and should provide project teams with the guiding principles of informal learning. What do we already know to be true about how individuals build on their existing knowledge structures; how do they integrate new information with existing information? It is possible that these knowledge structures are often fragmentary at best. How can the instructional designer maximize the learning experience? What variables should the designer have to take into account?

The tools for building theory may already exist in the work of others. Howard Gardner, Carol Dweck, and Lauren Resnick all have done seminal work in learning theory (to name a few). Carol Dweck's Cognitive Mediation Model was used to formulate the hypotheses for the Research Communications Ltd. study (1989) described in Chapter Two by Chen. The Cognitive Mediation Model is particularly appropriate for informal learning because teaching positive affect and self-confidence are considered appropriate goals for informal learning. And, in fact, using television to teach children consistently, minute-by-minute, that math is fun and that they can do math is much more feasible than teaching a math curriculum. One reason that the television medium does not lend itself well to teaching formal curricula is that children do not necessarily watch all programs in a series or watch in the same way. In contrast, any program and every program can teach the cognitive mediators necessary for success in mathematics or science in school. It is important if these models are used for conducting research, that the programs themselves are also using these models in their instructional design.

Developing New Methodologies

Much of the informal learning research field has focused on the stimulus rather than on the learner, focused on the stimulus in isolation of the context in which it functions, and ignored the complex interactions of the informal learning process with other learning sources, the passage of time, and influence of social mediators on learning. All of these factors are important.

Most respondents agreed that we do not have enough acceptable research models—that researchers need to explore new ways to achieve what has been referred to in the evaluation field as "authentic assessment." This means moving away from the traditional reliance on test scores and quantitative methods toward qualitative methods that get closer to the reality of the learning process. Mathematics, for example, is undergoing a shift toward measuring learning in terms of the problem-solving process rather than right answers with multiple-choice items. Tools that are becoming popular with this approach to evaluation include student portfolios, behavioral samples, and observation of group work. The *Square One TV* study conducted by Children's Television Workshop (CTW) cited in the chapter by Chen (CTW, 1991) developed problem-solving tasks to measure program impact. This procedure was labor intensive but provided significant insights into the show's potential impact on children ages 8 to 12.

Because so little is understood about these kinds of measures, the field showed a strong preference for more effective qualitative research measures and methods; many felt the anecdote tells the story of how informal learning works best. When little is known about a phenomenon, the first steps in anthropology are to observe, interpret, and look at how things work.

As the informal learning field grows, researchers need to demonstrate a much greater willingness to take risks in developing new models and methodologies. Traditional educational research methods that were developed for controlled laboratory and school settings are often inappropriate and insensitive to the dynamics of informal learning. Evaluators all too often defend their results as the truth. However, most experts agree that a failure to document "significant differences" is often a consequence of using research methods that are neither appropriate nor sufficiently sensitive to impacts. Few studies ever state at the end of the report that the research design or methodology was flawed. Yet trial and error is much needed in the field of informal learning research just as trial and error is needed in the development of the projects themselves. If these risks are not taken, the field of research will not grow. And if researchers are not much tougher on themselves about *why* significant differences were or were not detected, the field will stagnate.

Revisiting Longitudinal Research and Long-Term Effects

A primary concern of those interviewed was the need to develop models that will allow us to look at what happens with informal learning over time. On one level, informal learning research needs to determine what the cumulative impacts are of informal learning over a period of years. We recognize that informal learning begins at an early age and carries through adulthood, but how does the informal learning curve build across time? What is the maximum time to introduce different types of informal learning experiences? Is one type of experience more effective for younger children as compared with teenagers or adults?

On another level, we need to address the issue of measuring project-specific outcomes over time. What is the best time to look for impacts? As discussed in Chapter One, we need to give projects enough time to work, to have an effect. Yet most project-based research is completed before and soon after the project ends. There are logistical reasons for this. On television projects, for example, the project team disbands when the series is produced. On these projects, there is no audience for the research once the program airs. As one respondent put it, "We need more than a month and less than seven years. But we rarely get this right."

At this writing, one ongoing impact research effort is underway through the Chicago Museum of Science and Industry (Anderson and Roe, 1993). In this study, eight museums are looking at the impact of informal science learning experiences in their own institutions. No data were available at this writing but, hopefully, more efforts with this focus will be funded in the future.

Linkages Across Informal Learning Experiences

It is interesting to note that most informal learning media operate in isolation from one another. When one focuses on an informal learner such as our third-grader Cindy in Chapter One, we immediately see that she is experiencing a variety of informal learning experiences throughout her life. Yet, developers usually focus on their own media. They tend to pursue funding without giving consideration to other compatible resource institutions that are logical support systems. Professionals in these institutions rarely talk to one another. The end result is a highly fragmented set of informal learning experiences that do not systematically build on one another.

Even on the surface, this seems absurd. Imagine learning tennis by taking a lesson from one coach and only focusing on your forehand, another from a coach who teaches backhand, and yet another who teaches volley. No one would suggest such a system but that is how the informal science education "system" works.

This fact becomes even more absurd when you look at the institutions for informal learning within local communities where partnerships would seem logical and necessary. While some collaboration certainly does occur, local public television stations (which produce and distribute the television science programming) do not systematically launch community-wide outreach efforts in collaboration with local science museums. Yet, since these different media accomplish appreciably different types of learning experiences, one wonders why.

Some funders have recognized this lack of possible linkages and have begun to require some indication in proposals that other informal science learning activities will be undertaken. Because this is a relatively new phenomenon, it engenders some disgruntlement, but will ultimately result in a public that is better served by those who design, produce, and distribute informal learning.

The benefits of this cross-fertilization are obvious. Television is a cross-fertilization medium that is two-dimensional and shows process well, but museum exhibits can provide opportunities to interact directly with phenomena. Community-based projects offer yet another richness with its human interface. If projects could be constructed with the informal learner as the primary focus, media should then be selected and applied if and only if they serve the learner.

Interactions Between Formal and Informal Learning

Just as we need to seek linkages across informal learning experiences, the interconnections between formal and informal learning should be planned systematically. As technology expands to bring school-based experiences home, and home-based experiences to school, the lines between informal and formal learning will become blurred. This does not mean that out-of-school experiences should be extensions of the school curriculum, but there should be a healthy awareness and complementarity.

One respondent pointed out that the focus on the learner becomes even more important as the learner gains more and more control over his own learning; "This will be a messier environment to look at with less rigidly clustered pathways. The new technologies will take away rootedness in institutions and geography." The challenges of reaching those informal learners on a systematic basis will become increasingly complex as a plethora of educational television channels, home video products, interactive media, and computer networks fill the landscape. We know far too little about the learner as we move into this very complicated universe that is almost upon us.

Some were concerned about the crossover between formal and informal learning because informal learning practitioners felt that the formal system has essentially failed and that partnerships between the two could force informal learning to look like schools. This would be a serious problem since some of the most valuable informal learning goals, including fostering positive affect toward science, creating positive self-concept for doing science, and encouraging participation in science are primary, not secondary, goals for informal science learning efforts. Without these goals in view, students will continue to turn away from science and math in school. Without informal learning, the job of the schools is made even more daunting.

Future Directions for Informal Learning

When informal learning researchers and practitioners gather at conferences that tangentially relate to informal learning, the conversation usually comes around to the need for establishing informal learning as its own recognized field of endeavor. Professionals often identify the needs to encourage academic research, to establish research centers to train informal researchers, and to encourage scholarship in the field. Several practitioners interviewed for this chapter had attempted to establish research centers at universities.

Refereed research journals that are cross-disciplinary (i.e., represent museums, community-based projects, and television) are encouraged to build a literature in the field of informal science learning. Identifying specific professional meetings for informal science learning researchers and offering ongoing symposia with publications are also suggested. At the present time, the American Association for the Advancement of Science has a Committee for Public Understanding of Science and Technology. The establishment of a special interest group or division within the professional communications or educational research organizations is also warranted. A forum is needed where both academic researchers and formative and summative researchers and evaluators can share their work, participate in symposia, and discuss new ideas. As in the formal educational arena, it is always a challenge to get practitioners together with academics and to get academics into the real world.

Where Do We Go From Here? In conclusion, the research discipline in informal science learning is not very far along. We need to define it better, to broaden the vision for research. We need to gain a better understanding of how our informal learning projects work, but we cannot limit our view to isolated projects and their effects. We need to look at combinations of experience.

Working within this relatively new field of informal learning is exciting because it answers the most fundamental educational questions that this country faces. How are we going to excite and engage children and adults in the pursuit of science and make them understand its importance to their everyday lives? At one time, getting excited about learning was of little to no interest to our formal educational system. As that sentiment changes, schools and informal learning institutions may learn how to collaborate in the best interests of the public they serve.

Epilogue

This book was written by a group of individuals who value the research endeavor. But there are individuals who are not interested in their audience, have chosen not to listen to their audience, or have not followed that which has been learned from audience research on informal learning. And there will be those who will not be sympathetic to investigations devoted to project impacts and furthering an understanding of how informal science works. For those individuals, the following story is offered.

In his introduction to a chapter on how to find trout when fly fishing, Joe Humphreys (1981) shares this country tale.

A Pennsylvania farmer went into the general store and was sold a thermometer, which he took home and hung on a bush.

The next morning was the beginning of a cold snap that sent the mercury plunging. The first time he looked at the thermometer, the farmer counted ten frosty degrees above zero. The second morning the temperature had dipped to five above zero. The third morning there was a five-degree drop, and by the end of the week the thermometer recorded a frigid ten degrees below zero.

The following week the farmer went back to the store for supplies and the proprietor asked him how he liked the thermometer. "I had to smash the darn thing," was his reply.

"What on earth for?" asked the storekeeper.
"Well if I wouldn't have, we'd have all froze to death!"

The farmer, like our skeptics, didn't understand how a measurement tool can be helpful. There are those who would smash thermometers and those who would decree warm weather. Both would have missed the point. Neither would have caught trout.

References

Children's Television Workshop. (1991). *Children's problem-solving behavior and their attitudes toward mathematics.* New York.

Crane, Valerie. (1992). Listening to the audience: producer-audience communication. In *When science meets the public.* American Association for the Advancement of Science. pp. 21–32.

Humphreys, Joe. *Trout tactics.* Stackpole Books. 1981. p. 19.

Miller, Jonathan D. (1987). Scientific literacy in the United States. In *Communicating science to the public.* Ciba Foundation Conference. John Wiley & Sons: New York. pp. 19–40.

Museum of Science and Industry in Chicago. (1993). Museum impact and evaluation study: Roles and effect of the museum visit and ways of assessing them. Joyce Foundation.

Research Communications Ltd. (1989). *An examination of* Square One TV *as an informal math learning experience.* Dedham, MA.

St. John, Mark with the assistance of Deborah Perry. (1993). *Investments in informal science education: A framework for evaluation and research.* Inverness Research Associates. A paper prepared for the Association of Science and Technology Centers.

Annotated Bibliography

Part I: Impact Studies on Television

Touched by science: an exploratory study of children's learning from the second season of 3-2-1 Contact. MILTON CHEN, **1983.** Children's Television Workshop, New York.

Statement of problem/main purpose of study:

This study explored the range of cognitive and affective outcomes from frequent viewing of *3-2-1 Contact,* and provided feedback to CTW planning and decision-making for subsequent seasons. The Johnston & Luker study (1983) was also commissioned as an exploratory study of the second season of *3-2-1 Contact.*

Sample selected for the study:

Four mixed classrooms of 4th- through 6th-graders in a year-round school in Oakland, CA (N = 101) participated in the study. The sample contained 70% minority children and contained a diverse group of students in terms of family background and school performance.

Overview of methods used in study/procedures:

Participants viewed *3-2-1 Contact* programs daily, Monday through Friday, for a period of two to five weeks (10 to 25 shows). Ten students from each class participated in small-group interviews on a weekly basis. In addition, classes wrote brief essays on their learning from the programs. At the end of viewing, a post-viewing interest inventory was administered to all classes to examine relationships between viewing and scientific interest in 50 different topics.

Outcomes/major findings:

Based on the interviews and essays, learning from the series occurred in three general areas:

A. Children's Acquisition of Science Knowledge. Viewers were able to express newly acquired familiarity with a wide range of scientific concepts and phenomena.

B. Interest in Continued Learning. Children indicated that the programs motivated an interest that continued after viewing (e.g., asking further questions, reading a book, doing an experiment, making a model).

C. Shifts in Science Attitude. Some viewers were able to note shifts in their feelings about science, away from a perception of science as "boring" and requiring much study and lab work and towards a perception that science can be "fun." Accompanying this shift was a broadening of topics that fit under the label of "science" (such as scientific aspects of sports).

Television and children's problem-solving behavior: An evaluation of the effects of Square One TV. EVE HALL, EDWARD ESTY & SHALOM FISCH, 1990. *Journal of Mathematical Behavior and Children's Television Workshop.*

Statement of problem/main purpose of study:

Children's Television Workshop was interested in determining the effects of programs from the first two seasons of *Square One TV*, aimed at a home audience of 8- to 12-year-old children. Specifically, the study examined in detail the changes in children's attitudes towards mathematics and in their inclination to use problem-solving techniques as a result of sustained viewing of *Square One TV*.

Sample selected for the study:

The subjects for the study were 5th-graders in four public elementary schools in Corpus Christi, TX. This site was chosen because *Square One TV* had not been broadcast in this city prior to the study. A total of 48 children, 12 from each school, participated.

Overview of methods used in study/procedures:

The study utilized an experimental design with both pre- and post-tests. All 5th-graders in the two experimental schools viewed one half-hour episode of *Square One TV* each day. While viewing took place during the school day, it did not occur during mathematics classes, nor did teachers relate the program content to their usual mathematics instruction. The two control schools did not view *Square One TV.*

Subjects were individually interviewed and videotaped using a set of nonroutine mathematical problems. Their performance was measured by two scores: one involving the number and variety of problem-solving actions and heuristics used, the other involving the mathematical completeness and sophistication of their solutions. Subjects were also interviewed concerning their attitudes toward mathematics.

Outcomes/major findings:

Gains from pretest to posttest in both types of problem-solving scores were significant for the experimental group, and significantly greater for the experimental group than for the control group, indicating that frequent viewing of *Square One TV* can increase problem-solving performance in its target audience. In addition, this research showed that children's attitudes towards mathematics improved significantly as a result of watching the program.

The Eriksson Study: An exploratory study of viewing two weeks of the second season of 3-2-1 Contact. JEROME JOHNSTON AND RICHARD LUKER, 1983. Institute for Social Research, University of Michigan.

Statement of problem/main purpose of study:

This study attempted to conceptualize and collect data on the range of effects from viewing *3-2-1 Contact,* and to provide feedback to CTW planning and decision-making. The study by Chen (1983) was also conducted in a California site to provide additional data on the second season of the series.

Sample selected for the study:

Subjects were 192 fourth- and fifth-graders at the Eriksson Elementary School in Plymouth-Canton, Michigan. The student population at this school is predominantly white, from families with low- to upper-middle income.

Overview of methods used in study/procedures:

During the first week of the study, all students viewed the five shows in "Flight Week." During the second week, some viewed "Sports Week" and others viewed "Babies Week." Data were collected through essays and pre-/post-viewing interviews and questionnaires.

Outcomes/major findings:

After viewing, children's perceptions of scientists were beginning to change in desirable directions and many had retained a good deal of scientific information from the programs.

Appeal of the programs was high, especially for girls and for students who previously did not have much interest in science. Teachers had positive reactions to the program, noting that it had stimulated many questions from their students.

Final report: *Reading Rainbow Study.* NFO RESEARCH, INC., 1990. Reading Rainbow, Great Plains National, Lincoln, NE.

Statement of problem/main purpose of study:

This study was conducted to determine the reading habits and attitudes of families with 5- to 8-year-old children, as well as parents' familiarity with and attitudes toward *Reading Rainbow.*

Sample selected for the study:

A total of 1,098 questionnaires were sent to parents who were recruited with children between the ages of 5 and 8 years. The parents were previously recruited for NFO's consumer research panel. The sample was stratified for equal numbers of boys and girls, and equal numbers of 5-, 6-, 7-, and 8-year-old children. The sample was further stratified to match the U.S. Census profile of households with 5- to 8-year-olds for geographic region, age of head of household, annual household income, market size, and household size. A total of 707 completed questionnaires were received, for a response rate of 64%.

Overview of methods used in study/procedures:

Potential participants were mailed a four-page questionnaire designed to a cover a variety of topics, such as family reading habits, child's interest in reading, parental encouragement of reading, and parental attitude and child's response regarding *Reading Rainbow.*

Outcomes/major findings:

Most parents involved in this study read regularly and reported that their children enjoyed reading. Half responded that their children initiated reading by him/herself on a daily basis. Children who read regularly were likely to live in families that read books frequently and were encouraged to read by their parents. Boys were reported to read less frequently than girls did.

Seventy percent of parents were familiar with *Reading Rainbow.* Among those familiar with the series, over two-thirds encouraged their children to watch and over half reported that their child asked for books he or she had seen on the program. Children were more likely to ask for *Reading Rainbow* books if they visited the library regularly, liked to read both at school and for pleasure, and lived in a family that read regularly.

Science programmings and the audiences for public television: An evaluation of five programs in the NET Spectrum series. DAVID PROW-ITT, EDITOR, 1969. National Educational Television.

Statement of problem/main purpose of study:

This research was conducted to judge the impact on and acceptance by public television audiences of five half-hour programs broadcast in the *Spectrum* science series.

Sample selected for the study:

During the first phase of this research, 420 questionnaires were completed by teachers of high school and college science. These participants were solicited via mail, using a mailing list obtained from two sources.

During the second phase of this research, 505 college, high school and adult students in three different sites were shown the half-hour programs in small groups.

Overview of methods used in study/procedures:

Teachers of college and high school science were sent questionnaires and asked to ascertain the local air time for and view the program closest to their field of specialization. They were then asked to complete and return the questionnaire.

Science students viewed the films and completed questionnaires in small groups during a regular session of the class.

Outcomes/major findings:

A general significant knowledge gain was achieved by high school, college and adult learners. Learners and science teachers reported a satisfaction that science programs should be broadcast for the general public and that in addition to gaining a high acceptance rate by laymen, the programs encouraged and maintained interest in science.

Research findings for audience evaluation of "How About . . ." science reports. RESEARCH COMMUNICATIONS, LTD., 1987. **National Science Foundation.**

Statement of problem/main purpose of study:

The purpose of this research was to determine the level of awareness of the series *How About . . .*, the adequacy of topic coverage, the effectiveness of the news inserts, and the ways the inserts promote public understanding of science.

Sample selected for the study:

Phase 1: A sample of 98 commercial stations was selected from the total universe of current subscribers to the series. Distribution of the sample according to station affiliation, market rank, and region reflected that of the subscribership.

Phase 2: A sample of 210 viewers in ten sites nationwide was selected to reflect the viewing audience of the station newscast which carried the *How About . . .* science reports. All participants had viewed the programming in the preceding two weeks.

Overview of methods used in study/procedures:

Phase 1: The television stations were surveyed by telephone about a variety of issues surrounding the research.

Phase 2: Participants answered background questions on a questionnaire and then viewed a test videotape with fifteen *How About . . .* science reports. Following viewing, they responded to additional questions and participated in a group discussion.

Outcomes/major findings:

Most stations used the *How About . . .* inserts once a week or more, and reported a high level of satisfaction with all aspects of the spots. Many respondents felt that Don Herbert lent credibility to the spots, and reported that their audiences were very interested in the topics.

Substantial learning gains were identified following viewing. Topic interest varied by gender and age group. Many viewers expressed interest in obtaining more information on topics of particular interest.

A study of children's informal math learning. RESEARCH COMMUNICATIONS, LTD., 1989. **Research Communications, Ltd., Dedham, MA.**

Statement of problem/main purpose of study:

This study explored children's informal math learning experiences using *Square One TV.* Children's problem-solving processes as well as attitudes towards mathematics were assessed.

Sample selected for the study:

A total of 330 children ages 4 through 12 were recruited in 38 sites to participate in the study with one of their parents. Child/parent pairs were assigned to experimental or control groups with demographics carefully balanced across sites to ensure comparability across the two groups.

Overview of methods used in study/procedures:

All children were pretested on math problem-solving using eight problems selected from *Square One TV*. The experimental group of children and parents were then shown an episode of *Square One TV*, and children were asked to watch the program at home as often as possible during the second four weeks of the study. Parents in both the experimental and control groups maintained diaries of children's afternoon TV viewing to verify viewing frequency. At the end of the study, children were retested on the eight problems from the show.

Outcomes/major findings:

The findings of this study suggest that, by fostering a more positive and productive attitude toward math problem-solving situations, informal learning can lead to improved problem-solving performance.

FUTURES—*research summary*. RESEARCH COMMUNICATIONS, LTD., Spring, 1992. Foundation for Advancements in Science and Education (FASE), Los Angeles.

Statement of problem/main purpose of study:

This study looked at the long-term effects of the *FUTURES* series, under classroom conditions.

Sample selected for the study:

Seven junior high schools in four sites nationally participated in the study. A total of 88 students who saw *FUTURES* (users) were compared to 88 who had not seen the program (non-users).

Overview of methods used in study/procedures:

Half the participants viewed twelve different episodes of *FUTURES* in their math classes over the course of a semester while half did not view the series at all. Students were surveyed before any episodes were viewed, in the week after the last episode was viewed and a third time one month later. "User" responses were compared to "Non-User" responses.

Outcomes/major findings:

FUTURES was appealing to students (especially minorities) and had a positive, long-lasting impact on student attitudes. Students who viewed the program were more likely to agree that math and science were relevant to their careers than those who had not seen it. FUTURES positively affects attitudes toward hard work and the possibility of having a successful career.

Part II: Science Impact Studies on Science Museums, Aquariums, and Zoos

Characteristics of ideal museum exhibits. M.B. ALT & K.M. SHAW, 1984. *British Journal of Psychology,* **75,** 25–36.

Statement of the problem/main purpose of the study:

The purposes of these two studies were (1) to determine how visitors perceive exhibits and (2) to identify the most effective characteristics of exhibits from the visitor perspective. Alt and Shaw attempted to classify museum exhibits in terms of visitor perceptions, rather than from the view of the museum professional.

Sample selected for the study:

Natural History Museum (London) visitors over 16 years of age.

Overview of methods used in study/procedures:

The first study generated a list of descriptors used by visitors to describe exhibits. A total of 20 visitors were invited to study three exhibits, one at a time, and asked to tell the interviewer anything that struck them about each exhibit. Content analysis found that these responses fell into the following descriptive categories: attractiveness/noticeability, overall evaluation, clarity and ease of comprehension, evaluation of subject matter, required visitor response, emotional reaction, visual effect, appeal to different age groups. The total number of responses was reduced to 48 attributes with multiple items under each of the descriptive categories. For example, under the category "emotional reactions," items included "It's entertaining," "It's artistic," "It involves you," "It's good fun," and "I find it difficult to relate it to things or events in my life."

Using these attributes, a second sample of visitors was then asked to rate 45 exhibits in the second study. In addition, they were asked to state which of the attributes would apply to an ideal exhibit. Characteristics most strongly applied to the ideal exhibit included: "It makes the subject come to life," "You can understand the point/s it is making quickly," "There's something in it for all ages," "It's a memorable exhibit," and "It's above the average standard of exhibit in this exhibition." Strongly negative characteristics attributed to undesirable exhibits were: "It's badly placed—you wouldn't notice it easily," "It doesn't give enough information," "Your attention is distracted from it by other displays," and "It's confusing."

Outcomes/major findings:

This study suggests that the traditional educational principles which are used to design and critique museum exhibits need to be carefully reconsidered. In particular, it was noted that visitors rated most highly those exhibits which impart a short clear message, displayed in a vivid manner.

A perspective on field trips: Environmental effects on learning. J.D. BALLING AND J.H. FALK, 1981. *Curator*, 23/4, 229–240.

Statement of problem/main purpose of the study:

The authors attempted to gain insight into the ways in which the novelty of the museum environment affects visitors in school groups. The theory upon which this project was built suggests that humans have a basic need to feel safe and comfortable in their surroundings. It was hypothesized that a person in an unfamiliar museum environment is for this reason likely to learn a great deal about that environment, but very little of the educational content of the museum.

Sample selected for the study:

During four successive studies, samples of school children from various socioeconomic groups were selected.

Overview of methods used in study/procedures:

During the first study the behaviors of students from low socioeconomic-status families who had previously been relocated from urban housing projects to suburban housing projects, were compared with those of students from low socioeconomic-status families who had remained in urban housing projects. Students from each of these groups were matched by race, sex, IQ, grade in school, and age. All of them were pretested, and then taken to the Smithsonian Institution's Chesapeake Bay Center for Environmental Studies. There they participated in hands-on activities designed to teach them about ecological concepts. Following the trip, they were again tested.

In the second study students of a higher socioeconomic status were tested, and students from both urban and suburban neighborhoods were exposed to both novel and familiar settings. In this case, the familiar setting was represented by science activities within their own school yard, while the novel setting was represented by activities in a natural setting outside of the home community. Students were again pre- and post-tested for each setting.

In the third study the effects of the *degree* of novelty present in the setting were explored, as well as the number of examples of natural objects which were available for study (on the assumption that forests have more natural objects than do parks). Students from urban, suburban, and rural communities were taken to one of three settings: a small park in a large city, a park in a residential neighborhood, or a forest. As before, they were tested before and after participating in science activities.

For the final study predominantly white, middle class students of two different age groups were tested. Half of the children from each age group participated in science activities behind their school, and half of them went on a day-long field trip to a science center which they had not visited as a class before. In this case each student received a pre-test and two post-tests—one the day after the activities and the other a month later.

Outcomes/major findings:

Study One: While students in both groups learned about the wooded setting they had visited, only those who lived in the suburban housing project learned any of the conceptual information which had been presented during their visit. Providing confirmation of the effects of setting novelty was the observation that during their visit urban students had been much less attentive than were their counterparts.

Study Two: As before, students demonstrated concept learning in the familiar setting, but not in the unfamiliar setting.

Study Three: While urban students performed more poorly on tests than did those from either of the other groups, a significant effect was indicated according to place of residence. While there were some interaction effects, urban and suburban students tended to learn more in the forest (a setting which was somewhat novel, at least for suburban children, and which contained the largest number of examples of natural objects), while urban students learned most in the suburban setting (also somewhat novel, and a moderate number of examples of natural objects).

Study Four: All students showed significant learning on the day following the nature activities. Importantly, all of them scored nearly as high one month later, indicating long-term retention of their new knowledge.

In general, it was felt that these studies had shown that learning can take place on field trips, if the setting is carefully selected for the students who will participate. More specifically, the following conclusions were drawn: (1) Learning is highest when non-task behavior is lowest (non-task behavior includes daydreaming, horseplay, and "not attending to the educational task"), and (2) Non-task behavior is highest at the lowest and highest levels of novelty—in other words, optimal learning tends to occur in settings of moderate novelty.

Great expectations: Do museums know what visitors are doing? BEER, V., 1987. *Curator*, 30(3), 206–215.

Statement of the problem/main purpose of the study:

This study compared the discrepancies between the actual behaviors of museum visitors and the beliefs of museum staff members about those behaviors.

Sample selected for the study:

The behaviors of 1686 randomly selected visitors to ten museums in Los Angeles County were compared to the perceptions of twelve museum staff members.

Overview of methods used in study/procedures:

Visitors were observed during their visit, and then were asked to tell why they had come to the museum. Observations and interviews were focused on five "behavioral variables":

a. Goals for the museum visit
b. Time spent at displays
c. Exhibit materials
d. Use of the exhibit space
e. Evaluation of displays

Outcomes/major findings:

Museum staff members overestimated the time spent by visitors at displays. In reality, slightly more than ⅓ of all exhibits were viewed for more than 30 seconds, and almost half were skipped entirely. Stopping and time spent at exhibits appear to be partially dependent upon the characteristics of those exhibits. For example, while displays which had only text labels were skipped by 68% of the visitors, exhibits with some interactive element were skipped only 42% of the time. Overall, exhibits with a combination of text, audio, and other information-presentation techniques were skipped less than those with only one technique, but most of those failed to hold visitors for more than 30 seconds.

Staff members were correct in believing that visitors avoid reading labels. Even labels which accompanied interactive devices (expected to attract readers) tended to be ignored. It was suggested that audio-visual displays which demonstrate the use of interactives (rather than explanatory text) might be an effective means of encouraging their proper use.

Staff members must remember that not all visitors have education-related goals for their visit. With properly designed exhibits, even those who come to socialize or simply pass the time can also have a learning experience.

Brookfield Zoo's 'Flying Walk' exhibit. B. BIRNEY, 1988. *Environment and Behavior, 20/4, 416–434.*

Statement of the problem/main purpose of the study:

Chicago's Brookfield Zoo used a prototype to develop a new participative exhibit on the wing movements of birds. In a process known as *formative evaluation*, staff members attempted to determine whether the exhibit, as planned, (a) would be able to attract and hold visitor attention, (b) whether visitors would use it in the way which was intended, and (c) whether they would learn the concepts which were presented.

Sample selected for the study:

Data were gathered from randomly selected children and adults who visited the zoo, in four different phases.

Overview of methods used in study/procedures:

In order to pre-test visitors for their knowledge of the subject, an equal number of male and female child visitors were interviewed. Subjects were asked to imitate with their arms the wing movements of birds. Of those interviewed, 96% failed to correctly simulate wing movement.

Next, randomly selected groups of visitors were tracked as they moved past the prototype exhibit, to determine its effectiveness in attracting and holding the attention of groups of different sizes. It was found that for groups of two or more visitors, nearly half either used the exhibit, or watched someone else use it. It was also noted that those who participated in the exhibit and those who watched others participate stayed significantly longer in the overall exhibit hall than did those who simply walked past the exhibit.

In the third phase, individuals were observed as they approached the bird wing exhibit and stopped to participate. The behaviors of these visitors were continually recorded during the time in which they were actively engaged with the exhibit. It was found that a majority of visitors used the exhibit correctly, although very young children had some difficulty performing all the activities because of their size.

In the final phase, children leaving the exhibit were approached, and asked to imitate the wing movements of birds. In addition, they were measured to provide information of use in determining the most appropriate height at which to design the final version of the exhibit.

Outcomes/major findings:

Analysis of data found that the prototype exhibit had been successful in attracting and holding the attention of visitors, and that almost everyone used the exhibit in the way which was desired. In addition, it was found that the exhibit had been effective in teaching visitors the motion with which birds move their wings, as 45% of those interviewed after viewing the exhibit could correctly imitate those movements.

The effects of gallery changes on visitor behavior. S. BITGOOD & D. PATTERSON, 1993. *Environment and Behavior* (in press).

Statement of the problem/main purpose of the study:

The purpose of this study was to assess the impact of systematic changes in a small Egyptian Mummy gallery at the Anniston (Alabama) Museum of Natural History. Gallery changes included: adding labels to the wall, changing the physical characteristics of these labels (words per label, size of letters, size of label background, presence of an illustration, location of labels), and the introduction of a bronze bust reconstructed from a mummified individual. The study attempted to determine whether the introduction of new objects and elements in an exhibit functions in an attention-directing or an attention-distracting manner.

Sample selected for the study:

A total of 1125 randomly selected visitors to the Anniston Museum of Natural History.

Overview of methods used in study/procedures:

The study consisted of 10 phases: (1) baseline (containing no labels on walls); (2) one label (a label containing 150 words was placed on the wall); (3) three labels (the information from phase two was divided into three labels of 50 words each); (4) large font (the size of letters was increased from 18 to 36 font); (5) large background (the size of the label background was increased); (6) illustrations (a hieroglyphic illustration was placed on the labels); (7) near wall placement (the labels were placed on a wall closer to the mummy cases); (8) small font (the size of fonts were reduced from 36 to 18); (9) six labels (the number of labels and amount of information was increased from three 50-word to six 50-word labels); and (10) mummy bust (a reconstructed bronze bust of one of the mummy case inhabitants was placed in the gallery. A total of 1125 visitors were observed during the course of the study. Target behaviors included: total time in the exhibit area, time stopped viewing the various exhibit displays, time stopped reading labels.

Increases in the percentage of label readers occurred when (1) the 150-word label was divided into three labels; (2) the font size of letters were made larger; and (3) the labels were moved closer to the mummy cases. No change in the percentage of label readers was observed when: (1) the label background was increased; (2) illustrations were added to the label; and (3) the mummy bust was added to the gallery. As expected, a decrease in the percentage of readers occurred when the font size was decreased in phase 8.

Total time in the gallery between readers and nonreaders differed throughout the study. Time for nonreaders was similar to baseline (when no labels were present to read). Readers, on the other hand, averaged significantly more time in the gallery than nonreaders over and above the time necessary to read the labels. Time viewing the mummy cases and time viewing the x-ray display also increased compared with baseline and nonreaders. Adding the mummy bust in phase 10 increased total time in the gallery and time viewing the x-ray display for nonreaders such that they viewed the x-ray as long as readers. Readers and nonreaders viewed the mummy bust equally long.

Outcomes/major findings:

The results demonstrated that changes in a gallery can have a large impact on visitor behavior. Label reading was increased from about 12% to 56% with three changes (words per label, size of letters, and placement of labels). In addition, label reading was associated with substantial increases in viewing time, suggesting an attention-directing rather than an attention-competing function in the present study. Placement of the bronze bust did not significantly change the behavior of readers, but had a specific impact on nonreaders (i.e., increase of viewing time for the x-ray display).

Exhibit design and visitor behavior: Empirical relationships. S. BITGOOD, D. PATTERSON, & A. BENEFIELD, 1988. *Environment and Behavior*, 20/4, 474–491.

Statement of the problem/main purpose of the study:

The purpose of this study was to compare visitor behavior at similar exhibits across a variety of zoos in order to identify patterns that suggest general design principles.

Sample selected for the study:

Thirteen zoos from different geographical locations in the U.S. served as the settings for this study.

Overview of methods used in study/procedures:

Stopping and viewing time were observed as a function of animal activity, animal size, presence of an infant; and architectural characteristics such as sensory competition, proximity between animals and visitor, and physical features of the exhibit.

Outcomes/major findings:

Visitor behavior at similar exhibits was found to be consistent across geographical region and size of zoo. In addition, the following relationships between visitor behavior and zoo exhibits were observed: (1) active animals were correlated with twice as much viewing time as inactive animals; (2) viewing time was positively correlated with size of animals; (3) viewing time increased when infants were present; (4) visitor attention was decreased when exhibits were in visual competition with one another; (5) visitor attention increased when the animals were closer to the visitor; (6) visibility problems (low lighting levels, visual obstacles, visual screens) was correlated with less visitor attention; and (7) characteristics of the exhibit environment (simulated habitat) were correlated with stopping and viewing time.

Measuring the immeasurable: A pilot study of museum effectiveness. BORUN, M., 1977. Philadelphia: The Franklin Institute Science Museum & Planetarium.

Statement of the problem/main purpose of the study:

Borun states that in order to succeed as places of education, museums must elicit between visitors and exhibits a series of complex physical and mental interactions. To ensure that they achieve their goals in this way, museums should carefully evaluate and understand the experiences of their visitors. In order to provide feedback regarding existing exhibits, and to present a paradigm for this type of project, a one-year pilot study was conducted at the Franklin Institute Science Museum in Philadelphia.

Sample selected for the study:

1000 visitors to the Franklin Institute.

Overview of methods used in study/procedures:

Staff members of the Franklin Institute and other science museums were surveyed, in order to develop a list of institutional goals by which museum exhibits could be evaluated. These goals were then translated into questionnaires by which the museum and its exhibits could be evaluated. Testing was carried out in two phases.

In the first phase, different forms of this questionnaire were used to address concepts such as motivation for a museum visit, visitor interests, exhibit attendance, exhibit preference, and orientation. These forms were administered to both random visitors and teachers of visiting school groups. Some questionnaires were given before the visit, and some were given after, so that comparison of the two could provide clues to the effects of exposure to the museum. A total of 500 visitors were surveyed in this phase of testing.

In a second phase, some of the same instruments were used to evaluate the effects of changes which were made to the museum. Before- and after-visit test scores were again correlated, this time in an attempt to determine the effects of new orientation procedures. These new procedures involved the placement of an orientation desk at the entrance to the museum, where three brochures were made available to visitors. 500 individuals were again surveyed.

Outcomes/major findings:

Phase One:

With regard to visitors: Not surprisingly, during the school year school groups predominated on weekdays, while families were largely present on weekends. During the summer, family groups were more commonly seen, with an increase in out-of-town tourists. Teachers who were surveyed stated that they saw a museum visit as an important adjunct to formal education, primarily because of its perceived value in increasing interest among students.

With regard to learning: Overall, visitors were found to remember a great deal of the information appearing in the exhibits. School-age children were found to score especially well on tests.

With regard to attitude toward museums: Visitors were found to rate the museum lower in terms of positive feeling after a visit than before. It was thought that this may have been due in part to confusion resulting from construction projects which were ongoing at the time.

With regard to attitude toward science, scientists, and technology: While visitors overall showed a favorable attitude toward these concepts, attitudes measured after a visit were found to be significantly lower after a visit than before. Again, it was suggested that ongoing construction projects may have prompted this response.

Phase Two:

Two of the three orientation brochures were found to improve post-visit test scores for knowledge of exhibit content as well as attitude toward the museum. Visitors appeared to enjoy the quiz format of the brochures, and expressed favorable comments about their use.

General Findings:

Visitors were also found to prefer more complex exhibits (those with 30-40 displays per room), fewer background colors, and a greater number of participatory devices. In terms of the original goals of the study, the museum was found to be quite effective in achieving its educational goals, but somewhat less effective in stimulating curiosity, interest, and positive attitudes toward the themes which were presented.

The impact of a class visit to a participatory science museum exhibit and a classroom science lesson. BORUN, M. & FLEXER, B.K., 1984. Journal of Research in Science Teaching, 21/9, 863–873.

Statement of the problem/main purpose of the study:

Schools often use field trips to participatory museums as a means of providing hands-on experiences as an adjunct to traditional classroom teaching. Unfortunately little is known about the effectiveness of such experiences. This study assessed the cognitive and affective outcomes of school group visits to one museum. Three hypotheses were tested: (1) that students who visited an exhibit would score higher on a test of science knowledge than would students who had not, (2) that students who attended a lecture after visiting a museum would have higher test scores than would students who attended the lecture without a visit, and (3) that students would perceive their visit to be more enjoyable, interesting, and motivating than a lecture.

Sample selected for the study:

416 fifth- and sixth-grade visitors to the Franklin Institute Science Museum in Philadelphia.

Overview of methods used in study/procedures:

School groups were given a series of experiences within the museum, and then tested to determine the outcomes of those experiences. Students were divided into four experimental groups: a control group (students were tested before visiting the exhibits), an exhibit group (students were tested after viewing the exhibits), a lesson group (students were tested after hearing a lesson on the topic) and an exhibit/lesson group (students were tested after viewing the exhibits and hearing a lesson).

The exhibits used as the focal point of the study illustrated the concepts of levers, planes, and pulleys through hands-on manipulation.

The lesson was a 15-minute lecture on the same topic, conducted at the museum by a staff educator. One of two tests was given to each participant. Some were given a cognitive test which contained verbal questions (in multiple choice word format) and some were given a visual test (in multiple choice picture format). An affective questionnaire asked all students to rate their experience in terms of overall reaction, enjoyment, interest, perceived learning, and motivation for future learning.

Outcomes/major findings:

As hypothesized, students who viewed the exhibit scored higher on both verbal and visual tests than did those in the control group. However, the second hypothesis was not confirmed, as it was found that students who heard the lecture scored even higher, and approximately the same as those who had both seen the exhibit and heard the lecture. In response to the affective portion of the questionnaire, students reported overwhelmingly that the exhibits were more enjoyable and interesting than the lessons. Exhibits and lessons were rated nearly identically in terms of perceived learning.

Flexer and Borun suggested that the study had shown that while concept learning can take place in a museum, the more important finding was that the presentation of science information in an exciting way can stimulate interest and enthusiasm for the topic among students.

Naive knowledge and the design of science museum exhibits. BORUN, M., MASSEY, C., & LUTTER, T., 1992. Philadelphia, PA: Franklin Institute Science Museum.

Statement of the problem/main purpose of the study:

In this report, Borun et al. describe a 3-year study at the Franklin Institute Science Museum in Philadelphia. The study was centered on people's understandings and misunderstandings about the concepts of gravity and air pressure, and was intended to provide the foundation for new exhibits which would give people a more scientifically based interpretation.

This project was based on the philosophy that museum exhibits must do more than simply present accurate information; they must first correct any misconceptions which are held by people. Only then will visitors be able to assimilate new information. In order to do this successfully, exhibit designers must first have a clear understanding of those misconceptions—also known as *naive notions*. It had previously been confirmed that college and high school-level students, as well as elementary school children, maintain such incorrect assumptions. It had also been discovered that people tend to maintain naive notions until they are made aware of the flaws in their thinking. However, these ideas had never been carefully studied within the museum environment.

Sample selected for the study:

Museum visitors across a range of age levels were selected.

Overview of methods used in study/procedures:

For the initial phase of this study, museum visitors were shown an exhibit which illustrated the concept of gravity in action, and asked to explain what was taking place. Researchers found that very few could accurately explain the concept of gravity. It was also noted that approximately one-third of those interviewed believed that gravity was related to air pressure. This belief was indicated by statements such as, "It's in the air," and "Without air pressure things would not fall."

In the follow-up phase, visitors were shown a new exhibit which demonstrated that gravity was not dependent upon air pressure. After viewing it, a high proportion of visitors were able to correctly explain this exhibit, and many of them stated that the information presented in the exhibit was in opposition to what they had previously believed, and so had changed that belief. As one visitor stated, "If you look at that device, gravity doesn't need air." It was felt that the project at this point had demonstrated (1) that naive notions do exist among museum visitors, (2) that these notions do interfere with their ability to learn, but (3) that an exhibit which carefully addresses these incorrect notions is capable of teaching new information.

In the third phase of the study another common misconception about gravity was addressed. In this case, the belief was that gravity is in some way dependent upon the spinning of the earth. Another exhibit was designed and built in order to dispel this belief, and again visitors who had viewed the exhibit were interviewed. While the results in this case were less dramatic than before, a before and after comparison of visitors showed a significant improvement in accuracy.

The final phase of research focused on the development of the text labels which accompanied the spinning earth exhibit, described above. In an attempt to reduce further the number of visitors who maintained a belief that gravity was dependent upon the earth's rotation, a series of changes were made to the labels. Each new label was accompanied by a series of interviews to determine its relative effectiveness in dispelling naive notions.

Initially there were three goals for these labels: (1) to give operating instructions for the manipulative device, (2) to describe the action which occurs as a result of that manipulation, and (3) to teach abstract concepts which are illustrated by the device. However, none of the labels which were implemented were able to reduce the percentage of visitors with incorrect assumptions below 33%. Finally a new label was developed, with the following (slightly different) goals: (1) to give operating instructions, (2) to refer visitors to the device—for example, by asking them to use it to test or prove some concept, and (3) to offer the concept which is to be tested, as an explanation for why some phenomenon was observed. Post-visit interviews with these labels, which encouraged exploration of a concept rather than teaching a concept, reduced naive notions to 25%.

Outcomes/major findings:

In addition to the findings described above, the following principles were presented for exhibits which present complex or commonly misunderstood concepts:

(1) You must know how novice visitors think about the phenomenon you are planning to treat.

(2) You must analyze what you would like them to think.

(3) You must identify how the visitor's initial state and your desired final state compare with each other. If there is good correspondence, you are in luck. If you discover a very large gap, it might be best to re-think your goals. If there is some correspondence but also crucial areas of disagreement, take it slowly. Don't try to change too many things at once.

What's in a name? A study of the effectiveness of explanatory labels in a science museum. BORUN, M., & MILLER, M., 1980. Philadelphia: The Franklin Institute Science Museum & Planetarium.

Statement of the problem/main purpose of the study:

Borun and Miller conducted a series of five studies to examine the effectiveness of explanatory text labels in producing an educational museum experience. In particular, the project was intended to answer the following questions: (1) Do visitors choose to "study" the content of specific exhibits during their visit and, if so, what relationship does this have to other visitor behaviors?, and (2) What kind and length of explanatory label produces a significant increase in visitors' understanding and enjoyment of a display?

Sample selected for the study:

Visitors to the Franklin Institute Science Museum in Philadelphia.

Overview of methods used in study/procedures:

Study #1: *Whole Visit Study.* Randomly selected adults were tracked as they passed through the entire museum, and observers recorded the time they spent in each exhibit hall, time at each individual exhibit, whether or not they read labels, and whether they interacted in some observable way with exhibits.

Study #2: *Transfer of Misinformation.* This study focused on a thermo-conductor exhibit which required hands-on interaction from visitors. Observers at the exhibit recorded the occurrence and type of interactions which occurred.

Study #3: *Preliminary Labeling Study.* Observers at a gravity exhibit which had no labels observed conversations between adults and children, to determine the accuracy of the information offered by the adults.

Study #4: *Label Presence, Content, and Length Experiments.* Different label conditions were tested at the gravity exhibit to determine their relative effectiveness in educating visitors. During interviews, adults who had viewed the exhibit were asked to describe how a gravity tower worked, in order to determine their level of understanding of the basic concepts which were presented. In some cases their affective response to the exhibit was also tested.

Study #5: *Children's Explanatory Label Experiment.* Examined the relative effectiveness of four different labels on their ability to attract children as readers. These labels included one with an explanatory diagram, one with a colored border, one which contained a picture of a fictional character, and one which contained all three features.

Outcomes/major findings:

Study #1: Evidence suggested that during a visit to a museum, people will browse most of the exhibits, but will carefully study only certain exhibits which appeal to them. For those exhibits which they choose to study, visitors read an average of 68% of the labels which appear.

Study #2: In the thermoconductor exhibit, when no label was present only 6 percent of visitors participated in hands-on interaction. However, when the label was present 78 percent interacted.

Study #3: Observations of visitors to the unlabeled gravity exhibit revealed frequent exchanges of inaccurate information.

Study #4: The results indicated that explanatory labels can increase learning at an exhibit. Labels with historical information and everyday applications produced more gains in learning than did labels that described how the exhibit works, or the scientific principles by which it operates. It was also found that signs which presented three, four, or five topics resulted in lower test score gains than signs with one or two topics. In addition, as the number of topics was increased, the percentage of visitors reading the whole label decreased. Visitors tended not to read long labels, even though they expressed a preference for labels longer than those actually read.

Study #5: None of the label formats was significantly more effective in attracting young readers.

The behavior of family groups in science museums. J. DIAMOND, 1986. Curator, 29/2, 139–154.

Statement of the problem/main purpose of the study:

This study provided information on what family groups do during their visit to a museum.

Sample selected for the study:

Family groups of three or four people with at least one child were selected as participants.

Overview of methods used in study/procedures:

Observers requested permission to follow the family groups with one adult-child dyad selected as the focal subjects of that group. Behaviors and interactions between the focus dyads were recorded.

Outcomes/major findings:

The most common family group behavior was approaching an exhibit, manipulating it or observing someone manipulate it, and then withdrawing. Visitors were as likely to watch others manipulate an object as to manipulate one themselves. Parents often used "teaching" by verbal and nonverbal methods. Verbal methods included reading to a child. Nonverbal methods involved pointing to an object or pulling the child over to an exhibit, or handing an object to the child. The study suggests that both individual contact with the exhibit and social interaction among visitors are important aspects of museum learning.

Prototyping interactive exhibits on rocks and minerals. J. DIAMOND, 1991. Curator, 34/1, 5–17.

Statement of the problem/main purpose of the study:

Diamond suggests that the role of museums in a modern society should be to provide concrete knowledge of the world, rather than to try to discuss abstract concepts. In order to provide a paradigm for the development of exhibits which are successful in doing this, she describes the construction of a series of hands-on exhibits intended to teach visitors about minerals.

Sample selected for the study:

Visitors to exhibit prototypes at the San Diego Natural History Museum and the Cranbrook Institute of Science.

Overview of methods used in study/procedures:

Diamond describes an iterative process of design and evaluation, utilizing detailed exhibit prototypes. The process was coordinated by an independent, ad hoc group of exhibit designers and minerals experts, as well as a specialist in evaluation. This group placed special emphasis on carefully developing the interactive elements of the exhibits, rather than on the labels which would accompany them, in order to build intuitive exhibits through which even non-readers could learn.

Prototypes were initially evaluated through a process of observing and interviewing visitors. In the earlier stages of design, these observations and interviews were rather brief, with a goal of quickly gathering information on the types of behaviors which occurred, the amount of time spent interacting, and the level of understanding of the basic concepts which were being presented. After fine-tuning the prototypes to an acceptable level, a more serious evaluation process was initiated. At this stage a larger sample of visitors was observed, with a more rigorous system of documenting their precise interactions with each part of the exhibits.

At the same time, more detailed interviews were conducted, each of which began with general questions about what had been seen, and progressing to questions about how visitors would explain the key concepts to which they had been exposed. This second level of evaluation provided a firm empirical basis for the final design of the exhibits.

Outcomes/major findings:

Diamond concludes that while the process of building and evaluating prototypes can be both expensive and time-consuming, in the long run it can be a cost-effective means of building durable and educational exhibits.

California Academy of Sciences Discovery Room. J. DIAMOND, A. SMITH, & A. BOND, 1988. *Curator,* 31/3, 157–166.

Statement of the problem/main purpose of the study:

This study involved an ethological study of visitor group behavior to the Discovery Room of the California Academy of Sciences.

Sample selected for the study:

A total of 14,691 visitors were counted over a three-month period as they visited the California Academy of Sciences Discovery Room.

Overview of methods used in study/procedures:

The CAS Discovery Room allows visitors to touch and explore a collection of games, objects, and exhibits of natural history. Of the visitors counted, behavioral observations were made on a total of 62 groups and trackings of movement were recorded for 68 individual visitors while they were in the room. Also, during 75 half-hour periods, 2,417 visitors in the room were classified into age, sex, ethnic and social groups. Interviews with children and adults of 100 groups, 24 written surveys, and informal conversation were also used to obtain information.

Outcomes/major findings:

(a) The most popular objects were the discovery boxes (77% preference), followed by costumes (26%), puzzles (16%), the human skeleton (11%), phases-of-the-moon exhibit (11%), and mounted badger (8%).

(b) Favorite activities included touching (58%) and everything about the room (32%).

(c) Average time in the room was 18 minutes for parent-children groups; children interacted with an average of 19 objects.

(d) A child exploring the room alone was more likely to look but not touch. The presence of another individual greatly influenced the child's exploration. In fact, the child remained with the objects three times longer when accompanied by others.

(e) The most common behaviors in the room involved children handling and attempting to make the objects do something. Children sometimes invented new ways to use objects.

(f) Most of the visitors were there for their first time or accidentally came across the room.

(g) Adults tended to read labels aloud to children and suggested how objects could be explored.

(h) The adults were found to influence a child's exploration in two ways: (a) if the child felt timid, the presence of an adult helped the child feel more confident; (b) the adult presence caused the child to slow down long enough to develop interest in an object he/she would otherwise rush by.

(i) Children tended to be the initiators of activities and social contact with parents.

The authors argue that exhibits should be designed "to create a social environment as well as a physical structure." In addition, the following guidelines are suggested:

(a) Exhibits should encourage exploratory (open-ended) behavior.
(b) Exhibit objects should be real artifacts.
(c) Exhibit areas should be arranged so that they encourage social interaction.
(d) Staff should encourage exploratory learning.

Predicting visitor behavior. FALK, J., KORAN, J., DIERKING, L., & DREBLOW, L., 1986. *Curator,* **28/4, 249–257.**

Statement of the problem/main purpose of the study:

Falk and his colleagues suggest that explanations of visitor behavior can be divided into three distinct perspectives:

1. Exhibit perspective. According to this view, visitor behavior is controlled primarily by the characteristics of the exhibit (e.g., visual attractiveness, length of labels, intensity of illumination, size of letters). Attracting and holding power are the preferred measures of behavior.
2. Visitor perspective. Proposes that the major variables influencing visitor behavior are visitors' past experiences and interests. The authors suggest a metaphor that describes this view—"the visitor-as-shopper." Visitors often "shop" for a particular exhibit, making a beeline for that exhibit once they enter the museum. According to the visitor perspective, the best way to predict visitor behavior is to know more about the visitor. Attention and curiosity are the preferred measures from this perspective.
3. Setting perspective. This view assumes that behavior is primarily influenced by large-scale social and environmental factors, rather than by individual differences or the quality of the exhibits. The museum is viewed as a setting in which visitor behavior is well-defined and predictable. Variables such as proximity to an entrance or rest room, whether or not other people have stopped, and whether an exhibit is on the right or left side of a hallway are considered primary predictors of behavior.

The authors conducted a study to determine which of these perspectives provides the best solution to the question, "Can museum professionals in any way accurately predict the general behavior of visitors in their museum?"

Sample selected for the study:

60 visitors to the Florida State Museum of Natural History who were over the age of 16.

Overview of methods used in study/procedures:

In this study Falk and his colleagues obtrusively tracked visitors and recorded their attention to four elements of the museum environment: the exhibits, the setting, the visitors' own social group, other people, and self.

Outcomes/major findings:

It was found that visitors spent the first few minutes orienting themselves, and then became relatively focused on the exhibits. After 30 to 45 minutes, visitors appeared to become fatigued, and attention to exhibits decreased while attention to the setting increased. During this later time they moved more quickly past exhibits, becoming more selective about which ones they would stop to see. 15% of attention was constantly given to the visitors' own social group, while attention to self and attention to other people were relatively unknown.

Falk et al. stated that these findings were undoubtedly influenced by the way in which the research project was set up and the characteristics of the museum in which it was carried out. However, they did suggest that it had provided evidence that the setting perspective appeared most consistently to account for visitor behavior. In a more general sense, they suggested that it had shown that visitors behave in somewhat consistent ways, and that developing a better understanding of what controls these behaviors will lead to the provision of better museum experiences.

Development of scientific concepts through the use of interactive exhibits in museums. E. FEHER, & K. RICE, 1985. *Curator*, 28/1, 35–46.

Statement of the problem/main purpose of the study:

Feher & Rice suggest that the museum can be a valuable environment in which to study learning processes. As an example of such research, they studied the role of naive, or preconceived, notions on learning which takes place at museum exhibits.

Sample selected for the study:

This process utilized three sets of interviews with school children at two exhibits whose purpose was to explain principles of light and vision.

Overview of methods used in study/procedures:

Fundamental to the project was the idea that children must understand that visual phenomena are dependent upon interactions between the object which produces the effect, the person who serves as a receptor, and the light which passes from the object to the receptor. During the first set of interviews, students who viewed an exhibit were asked what they had seen, and how they would explain it. It was noted that the subjects consistently stated that what they saw was produced solely by the object.

Because it was important for them to understand their own active role as a receptor, during the second set of interviews, children were shown that the phenomenon which they had witnessed was partly dependent upon the way in which they viewed it, rather than solely on characteristics of the object. When these subjects were asked to explain what they had seen, it was found that they were much more likely to describe themselves as being a part of the phenomenon—"You see it." For the third set of interviews, children were asked to describe the nature of their role as a receptor.

Outcomes/major findings:

Overall, the authors identified four distinct types of explanations which children offered for the phenomena which they observed:

1. The effect is all in the object—no awareness of their own role as a receptor of visual stimuli.
2. The effect depends on the object and the receptor—without discussion of light as a separate unit with properties of its own.
3. The effect is due to the light going from the source to the object—belief that the phenomenon is primarily dependent upon characteristics of the light.
4. The effect involves the light, the object, and the receptor—awareness that the phenomenon is dependent upon the properties of three distinct, interacting elements.

Feher and Rice identified two important characteristics of museum exhibitions which appeared to greatly effect their success in producing a full understanding of physical phenomena. These characteristics are (1) that an exhibit be usable by the visitor in different ways, offering more varied exploration of a topic, and (2) that different exhibits explore the different aspects of the same phenomenon, providing a "richer learning atmosphere." In addition, the authors suggested that the study had shown the suitability of the museum environment as a place to study learning processes.

A naturalistic study of children's behavior in a free-choice learning environment. GOTTFRIED, J. L., 1979. Ph.D. Dissertation, University of California-Berkeley.

Statement of the problem/main purpose of the study:

This study examined children's behavior during school field trips to a biology discovery room. The purposes were to document children's exploratory and social behavior during the visit and to evaluate the educational impact of the field trip.

Sample selected for the study:

Subjects were fourth- through sixth-grade students that visited the Biology Discovery Room at the Lawrence Hall of Science.

Overview of methods used in study/procedures:

Several techniques were used to obtain data: surveys completed before and after the visit by both students and teachers; direct observation of behavior; focused interviews; map drawings of the room by students two weeks after the trip; sessions in which field trip participants taught peers by demonstrating what they had learned on the trip.

Outcomes/major findings:

The major results can be summarized as follows:

(a) Most teachers considered the field trip as enrichment rather than a continuation of classroom learning.

(b) Children seemed to progress through stages of exploratory activity: exploratory—touching and manipulating—analytic activities—creative activities.

(c) Children who were fearful of animals prior to their visit overcame their fear during the visit.

(d) Social influence via group size, observational learning and peer teaching-cooperation had an effect on exploratory behavior and learning.

(e) Pairs (or dyads) were the most commonly found group size, which also was correlated with more verbal and motor exploration.

(f) Two weeks after the trip, students were able to teach their peers the facts, concepts and skills learned on the field trip.

(g) Exhibits that could be manipulated were more effective in attracting attention than static displays.

Formative evaluation of exhibits at the British Museum. S. GRIGGS, 1981. *Curator,* 24/3, 189–202.

Statement of the problem/main purpose of the study:

When an exhibit is displayed, a primary goal of the museum is to communicate a message which informs and educates its visitors. The frequent failure of past exhibits to accomplish this has often been viewed as a failure on the part of visitors, who lack the knowledge or willingness to understand the message. The British Museum of Natural History, however, is one of the museums that are beginning to accept the responsibility for an exhibit's failure and devise a more pragmatic approach to the development of an exhibit. For them, the utilization of new evaluation techniques such as formative evaluation has been the key to developing a successful exhibit. Griggs describes two case studies in which formative evaluation was utilized in exhibit development at the British Museum of Natural History.

Sample selected for the study:

Museum staff members, and randomly selected visitors to the British Museum of Natural History

Overview of methods used in study/procedures:

For both case studies, the first stage of evaluation was a private showing of mocked-up exhibits to museum personnel, who provided written feedback on problems visitors may encounter. During the second stage, two visitor surveys were administered. During these surveys, visitors were asked to explain the storyline of the mock-ups in their own words (with the exhibit out of sight), and then to complete a multiple choice test in front of the exhibit.

Outcomes/major findings:

The results showed that the Hall of Human Biology had successfully communicated the important points, while the dinosaur exhibit did not (this was attributed to the fact that the material was obtuse). The actual percentages did go up for the dinosaur exhibit and was statistically significant but because formative evaluation is more criterion referenced a slight improvement of the dinosaur exhibit from 30% on the pre-test to 60% on the post-test was not viewed as a success because 40% still did not understand the message. Minor problems with the positioning of pictures in the biology exhibits were found during the questionnaire portion of the evaluation.

The evaluations proved to be very helpful in determining the success of the biology exhibit and in finding problems which would have led to a misinterpretation of a major point in the message. It also showed the ineffectiveness of the dinosaur exhibit to relay its message and that the implementation of a new idea was called for. Formative evaluation does not only help predict the probability of an exhibit's success, it saves the museum the time and money it would spend on correcting or rebuilding a bad exhibit. Last, it ensures that the museum's goal of communicating a message that informs and educates the visitor is achieved.

Orienting visitors within a thematic display. GRIGGS, S., 1983. *International Journal of Museum Management and Curatorship*, **2, 119–134.**

Statement of the problem/main purpose of the study:

Because museums tend to be large places, visitor orientation within the museum is a continual source of problems. Orientation can be viewed in two ways: topographically (in terms of the museum's physical layout) and conceptually (in terms of the subject matters). Each of these can in turn be addressed in variations of scale, from orientation throughout the entire museum building, to orientation within an individual exhibit.

It has been suggested that orientation problems can be reduced by two means: (1) by designing the museum environment in a legible, sensible way; and (2) by providing the appropriate "psychological set" by giving visitors the information they need to orient themselves. Griggs explored these concepts within a single exhibit area at the British Museum (Natural History).

Sample selected for the study:

Visitors to the British Museum (Natural History).

Overview of methods used in study/procedures:

A series of studies were carried out utilizing interviews and the tracking of visitors throughout the exhibit area.

Outcomes/major findings:

Griggs presents the following Recommendations for Effective Orientation:

- Orientation should be integrated into the process of developing displays and not tackled as an 'afterthought' once the exhibits have been developed.

- Orientation devices should be 'user-defined'. Instead of simply providing aids which have been developed by experts, they should be tested to ensure that visitors understand them.
- Design conventions should be made explicit. Visitors should be clearly told about any important organizational concepts or orientation systems.
- Detailed conceptual orientation needs to be tackled independently from topographical orientation. Because they are often quite different, the information which is provided to visitors as an orientation aid must be considered and developed separately from conceptual information.
- Topographical orientation should achieve two main objectives: to indicate to visitors the overall physical arrangement of the display, and to indicate the intended route through the display.
- At the entrance to a display topographical orientation can be provided by a grand plan. This plan must be large enough to attract attention, and simple enough to be understood quickly.
- This information needs to be reinforced throughout the display.
- Choice points should be identified during the development of the display and treated appropriately. If it is important for visitors to make particular choices, this information should be explicitly provided.
- Any breaks in the sequence of a display should be identified. When particular displays or information are unrelated to the primary theme of an exhibit area, this must be clearly indicated to visitors.
- The general aim of conceptual orientation is to produce in visitors the expectation of what the display is about and how it is organized conceptually.
- There are two main problems to be surmounted before conceptual orientation can be achieved: how to attract and hold visitors' attention to the necessary information, and how to communicate the information to them.
- Conceptual orientation should begin at the very start of a display.
- Conceptual orientation needs to be reinforced throughout a display.

Perceptions of traditional and new style exhibitions at the Natural History Museum, London. S.A. GRIGGS, 1990. *ILVS Review*, **1/2, 78–90.**

Statement of the problem/main purpose of the study:

Griggs reviews a series of research projects conducted at the Museum of Natural History in the 1970s and 80s. The impetus for this research was vociferous criticism of a series of new exhibits in the Hall of Human

Biology, which had been disparaged for lacking in sophistication because they presented fewer objects, and simpler written information than had traditionally been the case. More popular with critics were traditional exhibits which were based on the philosophy that "objects speak for themselves and that (the role of the exhibit developer) is simply to display objects and allow the objects to speak for themselves."

Sample selected for the study:

Visitors to the Hall of Human Biology.

Overview of methods used in study/procedures:

Through a series of surveys, it was found that the new exhibits were much more popular with visitors than were the traditional-style exhibits. Respondents initially indicated that at new exhibits they had studied the written information, remembered details of its physical layout, and had related the exhibit to their own lives. Follow-up study found that visitors attributed the following qualities to the new exhibits: they made learning easy, they were thought-provoking, they got information across clearly, they made the subject exciting, they had something for all ages, and they were designed with the ordinary visitor in mind. The older-style exhibits which offered more difficult information (and were generally praised by critics), were described by visitors as being difficult and unexciting.

Based on this initial evidence, it was decided to study the distinctions between old and new-style exhibits more fully. The first step was to develop, through hundreds of interviews, a lexicon of words and phrases which visitors use to characterize exhibits. Another sample of visitors was then asked to state whether it was important for each exhibit to have those characteristics.

Outcomes/major findings:

The result of these interviews was the identification of exhibit characteristics which were felt to be important, and which could be used to discriminate between different exhibits. It was found that new-style exhibits were consistently rated significantly higher in terms of a number of desirable characteristics, while old-style exhibits were generally rated more highly for undesirable characteristics.

For example, the newer exhibits were thought to use more familiar means of providing information, were easy for ordinary people to relate to, were more modern in design, and were appealing to children. On the other hand the older exhibits were thought to be more appealing to adults and to have more specimens, but for every other characteristic they were rated more poorly than the new exhibits.

Further examination of the answers given by those who preferred old and new-style exhibits identified three distinct groups of museum visitors: "the traditionalists, who preferred the old-style galleries (19% [of the total number of respondents]), the modernists, who preferred the new-style galleries (34%), [and] the remainder, who had no preference (47%)." Perhaps the most interesting distinction between these groups (considering the earlier criticism that the new exhibits would not appeal to more highly educated visitors) was that the "traditionalists" tended to be those with lower education levels, while those with more education preferred the newer, simpler exhibits.

The predictive validity of formative evaluation of exhibits. S. GRIGGS, & J. MANNING, 1983. **Museum Studies Journal,** 1/2, 31–41.

Statement of the problem/main purpose of the study:

This study tested the predictive validity of exhibit mock-ups in terms of their message, or storyline, and the perceived affective qualities of the exhibits from the visitor perspective. The primary research question was, "Are results from a two-dimensional poster-like mock-up predictive of the completed exhibits with varied three-dimensional displays employing a number of media?"

Sample selected for the study:

Visitors at the British Museum (Natural History).

Overview of methods used in study/procedures:

Two studies were conducted: (1) a comparison of the effectiveness of the story line for both mock-ups and finished exhibits; and (2) a comparison of the affective qualities of the mocked-up and finished exhibits.

Study 1: Storyline analysis

In the first study two samples of 30 visitors were selected. One group was shown the mock-up exhibits and then asked to recall the story as they understood it. Specific questions were also asked for further information. Interviews were tape recorded. The second sample of visitors were treated similarly, but they were shown the completed exhibits.

Study 2: Visitor perceptions

This study asked one group of visitors to indicate which characteristics describe mock-ups and another group was asked to indicate which characteristics describe the final exhibits. The results from these two groups were compared to see how closely the mock-ups and exhibits were perceived.

Outcomes/major findings:

Results from the first study indicated a strong relationship between the effectiveness of the mock-ups and the final exhibits.

Results from the second study showed that the mock-ups and the finished exhibits were perceived similarly in terms of the list of characteristics used in the study. For example, both mock-ups and final exhibits received high ratings on: (1) It involves you; (2) It deals with a complicated subject; (3) It deals with the subject better than textbooks do; (4) It teaches without being too serious; and (5) It makes a difficult subject easier.

Strategies for family learning in museums. HILKE, D.D., 1988. In S. Bitgood, A. Benefield, & D. Patterson (Eds.), *Visitor studies: Theory, research and practice, Volume 2.* Jacksonville, AL: Center for Social Design. pp. 120–125.

Statement of the problem/main purpose of the study:

Hilke has suggested that it would be worthwhile to study family learning in museums by focusing not simply on what the museum has to offer a family, but rather on the dynamics of the family group. In this project, she attempted to answer the following questions: (a) Do family visitors act as if they are trying to learn something about the museum exhibits? (b) What strategies do family visitors use to learn about museum visits? (c) What implications do these strategies have for exhibition design?

Sample selected for the study:

132 visitors in 53 family groups at a large natural history museum.

Overview of methods used in study/procedures:

Family members were observed individually, for approximately eight minutes each, as they visited the museum.

Outcomes/major findings:

Families acted as if they pursued an agenda to learn about the exhibits. Acquiring and exchanging information was the primary focus of family visitor attention. Eighty-six percent of what family visitors said or did was directed toward specific exhibits.

Hilke identified two basic strategies by which family members gathered information—personal strategies and cooperative strategies. Personal strategies (behaviors which are not dependent on others) included looking, reading, and manipulating the exhibit. These individual activities accounted for a majority of the information-seeking strategies which were observed, even though 90% of all behaviors occurred in the presence of other family members. This phenomenon indicated that family members come to a museum as a group, to pursue individual agendas.

Cooperative strategies (behaviors that are dependent on someone else) included asking for information, making statements, or answering in response. Because only 10% of all activities took place apart from other family members, it was observed that in spite of the individual agendas, museum-going remains very much a group activity. Information-sharing among family members tended to be in the form of spontaneous, rather than solicited statements between group members. Children and adults usually chose each other as partners in sharing activities, while it was less common for siblings to work together.

It was suggested that to succeed as places of learning, exhibit designers must understand and anticipate the goals and strategies of family groups. Hilke presents five questions which family members ask as they enter an exhibition hall: (a) What looks interesting in here? (b) What is there in here that I recognize? (c) What don't I understand in all this stuff? (d) How is all this stuff related to things that I already know or should know? (e) Is there something to do here?

The impact of interactive computer software on visitors' experiences: A case of study. D.D. HILKE, E.C. HENNINGS, & M. SPRINGUEL, 1988. *ILVS Review: A Journal of Visitor Behavior,* 1/1, 34–49.

Statement of the problem/main purpose of the study:

Prices for computer hardware and software are falling lower as their power increases, making the use of computers in museums appear more attractive. Unfortunately there remain many questions about the appropriateness and effectiveness of this new technology. This study attempted to provide answers to some of these questions.

Sample selected for the study:

Visitors over the age of twelve at two urban science centers.

Overview of methods used in study/procedures:

The Smithsonian Institution's Traveling Exhibition Service (SITES) developed an interactive computer program on the subject of lasers. This 15-minute program was intended to accompany a traveling exhibition which dealt with the history and application of lasers. The exhibit also contained wall-mounted panels with objects and text labels, display cases with objects, and various demonstrations of laser technology.

The computer program had two modules, one which explained how lasers worked, and the other which illustrated their current application in society. While these modules primarily presented information by a combination of animation and text, visitors were able to interact by controlling the pace of the presentation, and at certain points by inputting information.

Data were collected at the Maryland Science Center in Baltimore and the Discovery Center in Ft. Lauderdale, Florida on weekend days. Two experimental conditions were tested—one with the computer in operation ("computer on"), and one without the computer ("computer off"). During each of these conditions, a questionnaire was given to the first 25 visitors to leave the exhibition area. On these, individuals were asked to indicate the total time spent at the exhibit as well as time spent at the computers. In addition, they were tested to see if they could recognize important concepts from the exhibit, and to describe three uses for a laser. During each condition, visitors to the exhibit were also observed, and researchers recorded the number of visitors in each area of the exhibit, the demographics of those visitors, and their behaviors.

Outcomes/major findings:

When they were on, computers were found to attract more attention than any other part of the exhibit, and those who stopped to use the computers tended to stay for long periods of time. Visitors had little difficulty understanding the directions for using them, and appeared to have read all of the text which appeared. Parents were seen to participate actively with their children, often by reading the instructions while the children operated the computer.

This increased attention to computers reduced attention to other aspects of the exhibition. The "computer on" condition drew the most visitors from dynamic visual displays (which showed a 43% decrease in attraction), followed by traditional panels and cases (a 26% decrease) and other hands-on units (a 23% decrease). However, since the computers actually increased the number of visitors who attended the exhibit and the amount of time they spent there, it was felt that overall they did not have a negative effect on the rest of the exhibition. In addition, visitors were more likely to read text labels and look at other parts of the exhibition, and to discuss what they had seen among themselves when the computers were operating. Although it was found that visitors were no more successful in naming three applications of laser technology when the computers were in operation, they were more accurate in recognizing important concepts—even those which were not dealt with via computer.

It was suggested that computers can be an effective part of an exhibit, if they are "engaging," and if they "provide content that is integrated with exhibition themes."

Effects of questions on visitor reading behavior. K.D. Hɪʀsᴄʜɪ, ᴀɴᴅ C.G. Sᴄʀᴇᴠᴇɴ, **1988.** *ILVS Review: A Journal of Visitor Behavior,* 1/1, 50–61.

Statement of the problem/main purpose of the study:

Most museum exhibits are intended to educate visitors. In order for visitors to learn, exhibits generally require that visitors read informational labels. Unfortunately, studies have shown that few visitors take the time to read and comprehend what is written. While a number of methods have been proposed for getting visitors to read, their relative effectiveness is largely unknown. Hirschi and Screven attempted to provide some empirical evidence for one of these techniques by studying the use of labels which ask questions.

Sample selected for the study:

The study was conducted with family groups of visitors to five different exhibits at the Milwaukee Public Museum.

Overview of methods used in study/procedures:

Families were unobtrusively tracked as they passed through the five exhibit areas, and the amount of time they spent reading and viewing the exhibits was recorded. For each exhibit, Hirschi and Screven also developed a short add-on label which asked a question which could be answered by reading the information which was already present—for example, "Do Bears Hibernate?" and "Did Japanese Women Commit Harikari?"

These labels were then placed in the exhibits, and family groups were again timed as they passed through each exhibit area. Comparison of non-question reading times to question reading times showed an increase of over 1,300 percent—from a mean across all exhibits of 6.6 seconds without questions, to 95 seconds with questions. In addition, families who viewed the exhibits and had questions tended to stay and view the rest of the exhibit for a longer period of time.

Outcomes/major findings:

It was hypothesized that non-question labels received such poor reading times for reasons previously suggested by Screven (1986) including:

- The labels were poorly located and out of the natural line of sight
- Most of the labels were not located near the objects they described
- The labels were long and crowded
- The labels included technical terminology
- Most labels did not directly relate to the exhibit objects

While the study did not attempt to determine whether an increase in reading actually meant that visitors learned more, it was apparent that the use of questions worked well as a device to get them to stop and focus their attention. Beyond that, it was suggested that other factors relating to the informational content of the labels would determine learning—in particular whether or not labels provide information which is related to the knowledge and experiences of visitors.

Learning from developmental testing of exhibits. J.E. JARRETT, **1986.** **Curator, 29(4), 295–306.**

Statement of the problem/main purpose of the study:

Jarrett proposes that a process known as *developmental testing* can provide a practical means of constructing exhibits which successfully communicate with visitors. As its name suggests, developmental testing is generally conducted during the exhibit design process. In order to illustrate successful implementation of developmental testing, Jarrett describes the redesign of a series of related exhibits within the British Museum (Natural History).

Sample selected for the study:

Visitors to the British Museum (Natural History).

Overview of methods used in study/procedures:

Based on a preliminary survey, it was determined that very few visitors adequately understood the primary teaching points of three exhibits which dealt with the subject of genetic mutation. In an initial attempt to address the problem, new mock-up text labels were created and were placed within the existing exhibits. Developmental testing of the mock-ups was then carried out in two stages.

In the first stage, visitors who had viewed the exhibits were asked to describe what each was trying to explain. Brief follow-up questions also asked visitors to explain certain scientific concepts which were pre-sented in the exhibits, and which were felt by staff members to be critical to a proper understanding of mutation. Finally, subjects were asked whether they had found anything to be confusing, and whether they would suggest any changes be made to the text labels. Analysis of these responses showed that the three mocked-up exhibits were correctly interpreted by only 30%, 40%, and 70% of those interviewed, respec-tively.

The follow-up questions provided some detail as to why visitors failed to understand the exhibits. In general, it was found that the existing labels assumed a higher level of understanding of scientific concepts than was justified. In addition it was found that the labels also failed to address incorrect preconceived notions which were held by the interviewees. The results of this first stage of testing were used to guide the construction of a second series of mock-up labels. These new labels utilized more basic explanations of important concepts, employed simpler terminology, and attempted early in the sequence of exhibits to address incorrect notions of visitors before presenting more complicated information.

In the second stage of developmental testing, a new sample of visitors was then selected as they approached the area, and asked to participate in the study. After these individuals had viewed the exhibits, they were also interviewed within sight of the exhibits. This time, subjects were able to correctly explain the important concepts 70%, 75%, and 75% of the time, respectively. It was felt that this improvement provided a beneficial foundation for the permanent redesign of the three exhibits, and can serve to illustrate successful application of the developmental testing process.

Outcomes/major findings:

Changes made as a result of developmental testing improved the final design of all three exhibits.

Using modeling to direct attention. KORAN, J.J., KORAN, M.L., FOSTER, J.S., & DIERKING, L.D., 1988. *Curator*, 31/1, 36–42.

Statement of the problem/main purpose of the study:

Visitors are often confronted with unfamiliar situations within a museum. In such instances, they may act inappropriately, or fail to act at all. Either of these outcomes can reduce the effectiveness of exhibits—especially those with an interactive element. This study sought to provide information about the benefits of one relatively inexpensive method of dealing with such situations—the use of live behavior modeling.

Sample selected for the study:

Randomly selected visitors to the Florida State Museum.

Overview of methods used in study/procedures:

It was hypothesized that visitors who watched someone manipulate hands-on elements would be more likely to do so themselves. The study was conducted at two exhibits which offered visitors an opportunity to participate actively. The first of these offered objects to visitors for hands-on manipulation. The second exhibit offered headphone listening stations as part of a walk-through diorama.

To determine the rate of participation without modeling, visitors were randomly observed as they viewed each of these exhibits. Accomplices were then sent to the exhibits to model appropriate interaction with the objects and devices, and visitors were again observed.

Outcomes/major findings:

For the exhibit which simply presented objects, during both modeling and non-modeling conditions children were more likely to touch the objects than were adults. As hypothesized, the presence of models significantly increased touching behavior.

For the headphones exhibit without modeling, few visitors stopped to use the devices, and males were more likely to stop than females. It was also found that those who participated usually failed to use them properly. With the presence of models, however, visitor use of the headphone listening stations significantly increased, as did the amount of time visitors spent at the exhibit. In this case modeling appeared to be most effective for those over the age of 18, although no gender differences were found.

It was suggested that this study had shown that the use of live or videotaped models can inexpensively increase appropriate interaction with exhibits, and thereby improve the quality of visitor experience.

The relationship of age, sex, attention, and holding power with two types of science exhibits. J.J. KORAN, M.L. KORAN, & S.J. LONGINO, 1986. Curator, 29/3, 227–235.

Statement of the problem/main purpose of the study:

In an effort to provide some initial evidence of the value of hands-on exhibits, the investigators explored visitor encounters in the Object Gallery of the Florida State Museum. In this project, two research questions were addressed: (1) "Are visitors more attracted to exhibits they can manipulate?", and (2) "Is more or less time spent at an exhibit when visitors have an opportunity to manipulate objects?"

Sample selected for the study:

Visitors to the Object Gallery of the Florida State Museum.

Overview of methods used in study/procedures:

The focus of this project was an exhibit case filled with seashells, and three separate experimental conditions were utilized. In the first, the shells were presented under a Plexiglass cover. In the second, the shells appeared without a cover, and a sign indicated that touching was permitted. In the third condition, the shells remained uncovered, and a microscope was placed nearby along with instruction for its use in examining the shells. For each of these conditions, age, sex, and viewing time were unobtrusively recorded for each individual who approached the exhibit.

Outcomes/major findings:

The study found that (1) more visitors stopped at the exhibit when the shells were uncovered than when they were covered, (2) the amount of time spent at the exhibit was greater when the shells were uncovered than when they were covered, for visitors of all ages, (3) females spent consistently more time viewing the exhibit during all conditions than did males, (4) younger visitors tended to spend more time viewing the exhibit than did older visitors, and (5) while the percentage of visitors who stopped was nearly identical for the uncovered shells both with and without the microscope, viewing times for those who did stop were higher for both genders and all ages. Interestingly, in spite of the increased attention given to the exhibit in the open condition, only 38% of those who stopped actually picked up a shell, while for the open with microscope condition only 67% of those who stopped actually participated in this way. Other findings: (1) children would often return repeatedly to the open exhibit during the time they were in that room; (2) parents would often join their children when they showed an interest in examining the open shells; (3) visitors generally treated the open shells carefully and returned them to their compartment; and (4) while the microscope was present, smaller and unusually textured shells were most often chosen for examination, while larger shells and those near the edge of the compartment were more often chosen in the absence of a microscope.

The study showed that even low-cost manipulative exhibits can attract and hold the attention of visitors. This appeared to be so regardless of whether or not a given visitor actually picks up a shell— indicating that manipulatable objects have an attraction regardless of a given individual's inclination to participate. It was felt that the behaviors which were observed with the hands-on exhibits—stopping, touching, and manipulating—are indicative of curiosity within visitors, and that curiosity may be an important precursor to education.

Oh, yes, they do: How museum visitors read and interact with exhibit texts. MCMANUS, P.M., 1989. *Curator*, 32/3, 174–189.

Statement of the problem/main purpose of the study:

It is often stated that museum visitors do not read text labels. McManus suggests, however, that the evidence upon which these statements are based is primarily obtained through flawed methods of observation. In order to study visitor reading more carefully, she compared visual observations of reading with verbal statements which were recorded at exhibits.

Sample selected for the study:

1,571 individuals in 641 visitor groups at five exhibits in the British Museum (Natural History).

Overview of methods used in study/procedures:

Visitor reading was recorded at five different sites throughout the museum and compared with unobtrusively made recordings of the same individuals.

Outcomes/major findings:

While a high percentage of visitor groups were by observation recorded as not having read labels, analysis of recordings showed that most groups had at least one individual who not only read, but exhibited "text-echo." Text-echo is the term given to the repetition of text information in conversations among visitors. These individuals were often recorded paraphrasing text, or repeating phrases from the text. Text-echo was measured in over 70% of the transcripts taken from four different exhibits. Of those groups whose transcripts contained text-echo, over 25% were not observed reading.

Three phenomena relating to visitor interaction with text labels were noted in this study: (1) Visitors seem to feel as if someone is talking to them through the label; (2) Visitors often talk back to this "someone"; (3) Visitors tend to limit conversation to the topic set by the label. The implication of these findings is to suggest that a conversation continually takes place between the exhibit designers and the visitors, and the nature of this conversation should be carefully considered during the design process.

Based on this information, McManus proposes six principles that illustrate characteristics which visitors require of exhibit labels.

1. Satisfaction of the visitor's interrogative framework: labels must provide a hierarchy of information which satisfies visitors' natural curiosity. For example, a label must first give the topic, then tell what is being said about the topic, and so on.
2. Adherence to the topic: designers must avoid shifts in the topic after it has been established.
3. Establishing orientation to the topic—the conversational frame: McManus suggests that it may be more useful to simply provide introductory material on the topic, rather than trying to make connections between the topic and the everyday life of the visitor.
4. Readability and conciseness of text: Labels should provide information which can be read and absorbed quickly.
5. An appropriate conversational tone: designers should avoid closed questions such as, "This is a predator, isn't it?" Instead, provide questions which require some exploration or interpersonal interaction to find a solution.
6. An appropriate social tone: labels should avoid authoritarian directives (e.g. "Look at this!").

Impact of exhibit type on knowledge gain, attitudes, and behavior. B. PEART, 1984. Curator, 27/3, 220–227.

Statement of the problem/main purpose of the study:

This study compared the effectiveness of several types of exhibits on visitors' knowledge gain and attitudinal change, the exhibits' attracting power and holding power, and the amount of interaction which takes place between visitors and exhibits.

Sample selected for the study:

First-time visitors to the "Living Land-Living Sea" Gallery of the British Columbia Provincial Museum in Victoria, Canada.

Overview of methods used in study/procedures:

The *Object* condition presented the exhibit object by itself without label. The *Word* condition consisted of a label only. The *Picture* condition used a picture of the object and a label. The *Standard* condition included both object and label. Finally, the *Sound* condition included the object, the label, and sound.

Outcomes/major findings:

The following results were found:

Condition	Knowledge	Attract Power	Holding Power
Control	38.4%	—	—
Object	39.0	55%	.47
Word	59.5	23	.43
Picture	68.6	23	.69
Standard	63.4	61	.77
Sound	71.4	91	.79

It is clear from these results that presenting the object without a label resulted in little or no learning compared with the control subjects who did not view the exhibit. Note also the differences in attracting power between the Picture condition (23%) and the conditions that included the object itself (55% for Object, 61% for Standard, and 91% for Sound). Finally, with respect to the attracting power, adding sound to the object and label increased the attracting power by 30 percentage points. The holding power was greatest for the Standard and Sound conditions. Peart also found a high correlation between the holding power and the knowledge gain ($r = .800$).

Analysis of a natural history exhibit: Are dioramas the answer? PEART B., & KOOL, R., 1988. *International Journal of Museum Management & Curatorship*, **7**, 117–128.

Statement of the problem/main purpose of the study:

The purpose of this study was to determine the effectiveness of an exhibition gallery at the Royal British Columbia Museum. In particular, museum staff members wished to answer the following questions:

1. Does a visit to the gallery result in knowledge gain?
2. Does a visit to the gallery change attitudes?
3. If knowledge gain and attitudinal change do occur, are certain exhibit types more effective than others?
4. Does exhibit type affect visitor flow patterns and behaviors?

Sample selected for the study:

112 randomly selected first-time visitors to the RBCM.

Overview of methods used in study/procedures:

The focus of the study was a gallery entitled *Living Land Living Sea*. This gallery contained both open and closed dioramas, as well as small didactic exhibits. The 46 exhibits of which it was comprised were of two basic types: concrete and abstract. Concrete (or realistic) exhibits were 3-dimensional and contained objects, while abstract exhibits were of a flat, 2-dimensional format. Later in the study this index of concreteness was changed to a more continuous scale which was based on a number of factors including the size of an exhibit, whether the exhibit was open (without glass) or closed, and whether it contained graphics, sound, motion, smells, specimens or models. Larger exhibits were assumed to be more concrete, and those which stimulated other senses were also felt to increase the concreteness of an exhibit.

Questionnaires were administered to 56 visitors immediately upon leaving the Living Land Living Sea Gallery, and to another sample of 56 who had not yet visited the gallery. In addition, visitors were unobtrusively tracked as they moved through the gallery.

Outcomes/major findings:

In regard to the first research question, comparison of pre-visit and post-visit questionnaires showed that learning did take place at the exhibit, although it was noted that this increase in knowledge was of relatively small magnitude (a 12% increase). However, no significant change in attitude was found between visitors who had, and those who had not viewed the gallery (the second research question). To the third question, exhibits judged to be more concrete were found to be more successful in attracting and holding the attention of visitors—however, this appeared not to translate to significant changes in knowledge or attitude.

Finally, analysis of observable behaviors found that traffic flow was influenced by exhibit type. Unfortunately other factors such as architecture, crowding, and exhibit location also were critical determinants of traffic patterns, which generally bore little relationship to those intended by designers. In addition it was noted that while approximately 74½ minutes were required to adequately view the entire gallery, visitors actually spent an average of just under 14 minutes in the process. Of that time, 35 percent was spent doing things other than looking at exhibits. In judging the overall success of the Living Land Living Sea Gallery, Peart and Kool suggested that although visitors appeared to enjoy themselves, museum professionals would have to be critical of its effectiveness.

A naturalistic study of visitors at an interactive mini-zoo. S. ROSEN-
FELD, & A. TERKEL, 1982. *Curator,* 25/3, 187–212.

Statement of the problem/main purpose of the study:

This study investigated the activities of family groups in an interactive
mini-zoo at the Lawrence Hall of Science in Berkeley.

Sample selected for the study:

Twenty-three randomly selected groups of visitors were observed and
16 groups were interviewed.

Overview of methods used in study/procedures:

The mini-zoo included six animal exhibits and seven Zoo Game
exhibits which were intended to familiarize visitors with the concept of
animal adaptation to the environment. Animals which were present
included a llama, a donkey, a goat, a hen and chicks, a chameleon, a
macaw, an opossum, and an anaconda. Zoo Games offered visitors a
chance to compare their physical features and skills with those of
animals. These games were titled How far can you jump?, How tall are
you?, How fast can you run?, How fast does your heart beat?, Sunflower
Seed Eating Contest, Animal Detective Game, and Draw an Animal.

Each family group was observed as they passed through the exhibit
area. For each group, observers recorded the following: (1) the sequence
in which the exhibits were visited, (2) the level of interaction with each
animal or game, (3) time spent per exhibit, (4) total time at the mini-zoo,
(5) group members who initiated the group's progression through the
area, and (6) verbal comments between group members. At the conclusion of their tour, some groups were also interviewed. In these interviews, group members were asked sentence fill-in questions (e.g. "What I
liked best about the animal fair was . . ."), as well as open-ended
questions (e.g. "What did you like best about the animal fair?"), and
picture-stimulus questions, for which children were shown pictures of
other children and asked to describe what they might be saying about the
exhibits.

Outcomes/major findings:

Based on these observations and interviews, the following conclusions were drawn:

a. Visits were primarily child-oriented. Adults wanted children to
   enjoy themselves.

b. Children interacted more with animals and the zoo game; adults were more likely to spend more time reading labels.

c. Adults tended to focus children's attention on doing the activities correctly, and on interpreting the activities for them.

d. Children controlled the pattern and pace of the visit in 69% of groups. But once a family group stopped, adults tended to direct the action.

e. Almost 40% of the overheard comments related to group management, or orientation.

f. The primary styles of group orientation within an exhibit included "casing out" (one group member gets an overview of the exhibit and then guides the others), "wandering" (the most frequent style), and "specific focus" (group members had come to see a particular animal).

g. The animal exhibit and zoo games with the most interactive elements were the most popular ones.

The effectiveness of guidance devices on visitor learning. Screven, C.G., 1975. *Curator,* **18/3, 219–243.**

Statement of the problem/main purpose of the study:

Visitors to an exhibit of iridescent glass at the Smithsonian Institution were studied in order to develop an understanding of the effectiveness of guidance techniques in producing an educational museum experience. This project was based on the theory that a number of variables interact to produce learning in visitors. These variables can include (1) the unique characteristics of the visitors (e.g. knowledge, interests, expectations, etc.), (2) the physical layout of the exhibit (e.g. lighting, space, colors, labels, and objects), and (3) the way in which visitors are allowed to interact with the exhibit (by pressing buttons, moving panels, etc.).

Screven suggests that motivational (or guidance) devices which are external to the exhibit can encourage visitors to think and learn by focusing attention on important aspects of a display. Such aids can include hand-held items such as audio tapes, question-and-answer cards, explanatory programs, and self-scoring punchboards. For this study, variations of these adjunct devices were compared, in order to assess their relative effectiveness.

Sample selected for the study:

736 museum visitors fourteen years of age and older.

Overview of methods used in study/procedures:

Visitors were both allowed to volunteer, and approached and asked to participate by viewing an exhibit with the aid of a particular guidance device. Among those items tested with the first sample of visitors were (1) a portable self-scoring punchboard which provided information not available in the exhibit and also asked questions about that information for visitors to answer, while providing immediate feedback on the correctness of responses, (2) an audio tape which provided additional information and questions, and automatically stopped to allow visitors to consider the questions, (3) an audio tape which provided additional information and questions, but merely paused for a few seconds to allow visitors to consider questions, (4) an audio tape which presented information but did not ask questions, and (5) a question-and-answer booklet. In addition, the exhibit was tested in its original format (with no guidance devices), and with the presence of two types of supplementary text labels that were placed within exhibits and tested as well. These included (1) labels which provided additional information about objects, and (2) labels which not only provided additional information, but asked questions of visitors. Each person was then asked how well they liked the device they had used, and was given a series of questions about basic concepts which had been presented.

A different sample of visitors was allowed to view the Frederick Carder Glass Exhibit with the aid of each of these guidance devices. In order to provide a baseline comparison for these conditions, one group of visitors was also tested without having seen the exhibit, and another was tested after reading the question-and-answer booklet but without having seen the exhibit. To determine the effectiveness of the different guidance devices, scores for each condition were compared to the scores which were obtained when the exhibits were viewed in their natural state.

Outcomes/major findings:

The average scores of visitors who were tested were higher when using any of the guidance devices than when viewing the exhibit in its original state. However, scores were considerably higher (indicative of a greater amount of learning) for all of the hand-held items than for the labels which were added to the exhibit.

Audio tapes were found to be both the most popular form of adjunct device, and the means by which the greatest learning was achieved. Tapes which asked questions were particularly effective, although there were no significant differences between the tapes which stopped automatically, and those which merely paused. While booklets and self-scoring devices were also effective, they received lower scores than did audio tapes, and it was also noted that fewer visitors took the time to actually complete the booklets than to complete the tapes. Screven felt that the text of the booklets, while identical to its taped counterparts, was more intimidating to potential readers, and required more time and effort for completion.

Understanding processes of informal education: A naturalistic study of visitors to a public aquarium. TAYLOR, S.M., 1986. Ph.D. Dissertation. University of California, Berkeley.

Statement of the problem/main purpose of the study:

This study examined the relationship between learning and visitor behavior in the Steinhart Aquarium in San Francisco.

Sample selected for the study:

Family groups were randomly selected from among visitors to the California Academy of Science.

Overview of methods used in study/procedures:

Obtrusive methods of observation (visitors were asked to participate prior to being observed) were employed to record patterns of traffic flow through the aquarium and time spent in each area of the facility. Various techniques also were used to elicit questions about the exhibits from visitors. Ethnographic analysis was used to record and evaluate visitors' thoughts, feelings, and motivations as they moved through the aquarium.

Outcomes/major findings:

Taylor drew the following conclusions after analyzing this data:

a. Traffic patterns in the aquarium followed predictable patterns. Visitors tended to turn right and follow a counterclockwise pattern through the facility.
b. There was definite evidence of "museum fatigue," or a decreasing time at each exhibit as the visit progresses.
c. Most visitors expressed a desire to see the whole aquarium, but were unwilling to backtrack, and thus, they missed some sections.
d. Visitors' questions about the displays indicated an interest in concrete, literal, and visually verifiable aspects of the displays.
e. Visitors' conversations centered on the theme of familiarity. They discussed the exhibits in terms of things with which they had some previous experience.

f. Most of the visitors were constrained for time during their visit to the aquarium.

g. Visitors sought out interactions with the animals.

h. The most common use of the labels on the displays was for identification of the fish on display. Very few groups read the label text in depth.

i. Misconceptions in label reading were frequent.

j. Parents frequently translated the label text into simpler language for their young children.

k. Information flows to aquarium visitors were from three sources: visitor conversations, direct observations of the displays, and from the label texts.

Matching visitor learning style with exhibit type. C. VANCE & D. SCHROEDER, 1991. A. Benefield, S. Bitgood, & H. Shettel (Eds.), *Visitor studies: Theory, research, & practice, Volume 4.* Jacksonville, AL: Center for Social Design. pp. 185–200.

Statement of the problem/main purpose of the study:

Vance and Schroeder hypothesized that studies on learning style might help museum professionals design more effective exhibits. It was felt that this literature might offer new insight into the effects of psychographic characteristics of visitors such as values, attitudes, perceptions, interests, expectancies, and satisfactions. In particular, they focused on the use of the Myers-Briggs Type Indicator (MBTI), an instrument currently used to identify learning style. The MBTI is a forced-choice, self-report personality inventory. Natural preferences are broken into two poles on each of four indices: (a) E/I (extroverted/introverted); (b) S/N (Sensing/Intuitive); (c) T/F (Thinking/Feeling); and (d) J/P (Judging/Perceptive).

Four indices yield sixteen possible combinations called "types," signified by 4 letters of preference (e.g., ESTJ, INFP). The Sensing-Intuitive preference seems to be an important one for learning. Sensing individuals are more interested in immediate data taken in through their senses. Intuitive types are more interested in perceiving the relationships, meanings, and possibilities suggested by experience. Studies suggest that intuitive types are outnumbered by the sensing types.

Sample selected for the study:

Visitors to the Milwaukee Public Museum.

Overview of methods used in study/procedures:

Subjects were visitors (188 males and 212 females all over the age of 18) to the "Rain Forest: Exploring Life on Earth" exhibit at the Milwaukee Public Museum. Baseline, control, and two experimental conditions each had 100 subjects. Subjects were asked to participate after having been randomly chosen either before or after they had viewed the exhibit.

As they entered the museum, the control group of 100 subjects were asked to fill out an 11-item questionnaire on target exhibits and complete the MBTI. This was done prior to viewing the exhibits of interest. The control group reflected visitors' entering knowledge. In the other three conditions (baseline and two experimental conditions), 100 visitors in each condition were asked to fill out the questionnaire and MBTI either before of after viewing the exhibits. Observers used both cued and non-cued techniques. In the cued technique, 50 visitors were told before they entered an exhibit that they would be tracked and observed during their inspection of each exhibit. In the non-cued technique, visitors were approached after viewing an exhibit and asked to participate in a research project.

Labels were constructed for the experimental phases of the study. The first experimental condition included "intuitive" labels. These were created to appeal to the intuitive learners who are interested in reading and problem-solving. The second experimental conditions used labels that were "sensing" in nature for those learners who directly apply their five senses to each exhibit. The labels included concrete facts that explained why something is the way it is, rather than problem-solving or hypothesizing about the answer. Since there was no difference between the cued and non-cued visitors, these groups were combined for the reporting results.

Outcomes/major findings:

More than a third of respondents were categorized as sensing and 65.2% were intuitive. This was counter to the results of Myers and McCaulley (1985) who reported 75% of the general population are sensing, while 25% are intuitive. Analyses were performed to measure the differences, if any, between sensing and intuitive types on the rain forest questionnaire. There were no differences in performance during baseline or control. However, intuitive visitors performed better when intuitive labels were in place and sensing learners performed better when sensing labels were used.

Using time as a variable, there was no difference in amount of time spent at each exhibit during baseline of sensing and intuitive visitors. In the first experimental condition, with intuitive labels in place, intuitive learners spent more time than sensing learners. In the second condition, no difference was observed.

Analyses were computed to determine how sensing and intuitive visitors differed in response to the rain forest questionnaire. Sensing visitors scored significantly higher in the 2nd condition than in baseline, control and the 1st condition. There were no differences in baseline, control and the first experimental condition. Intuitive visitors in the 1st condition did score higher than in the baseline, control, and the 2nd condition. No differences were found in baseline, control and the 2nd condition.

The results further emphasized the importance of matching visitor learning style with informational qualities of exhibits. Identifying these different learning styles is an important step toward taking advantage of the powerful educational opportunities available in the museum setting.

Part III: Impact Studies For Community-Based Efforts

Final report on the effectiveness of intersegmental student prepara-
tion programs. CALIFORNIA POSTSECONDARY EDUCATION COMMISSION, 1992.
California Postsecondary Education Commission.

Statement of the problem/main purpose of the study:

The studies of MESA included in this volume attempt to show that
African-American, Mexican-American, and Native American students
who participate in MESA classes and activities during middle school
and/or high school are more likely to take advanced mathematics and
science courses and complete other requirements for college entrance
than are their peers.

Sample selected for the study:

One study compared MESA participants graduating from high school
in 1990 to all California graduates that same year. A second study
(Appendix H) sampled students in grades 7–12 who had participated in
MESA for at least one year as of April 1991.

Overview of methods used in study/procedures:

The first study was a comparison of the percentage of MESA participants *completing* advanced mathematics, chemistry, or physics courses
in 1990 with that of California high school students *enrolled* in those
courses in 1989 and the percentage of each group taking the SAT as
seniors in 1990. In the study in Appendix H, a four-part questionnaire
was sent to 408 MESA participants (10% of the potential pool) chosen
randomly; it was returned by 241 (59%).

Outcomes/major findings:

MESA participants were more than twice as likely to *complete* advanced mathematics or chemistry courses, and over four times as likely
to *complete* physics courses, as all California students were to *enroll* in
these courses. They were three and a half times as likely to complete
advanced mathematics courses and nearly eight times as likely to complete physics courses as all African-American and Latino students were
to enroll in these courses. The ratio of MESA seniors to all California and
African-American seniors taking the SAT was greater than 3:2; for MESA
and Latino seniors the ratio was greater than 2:1.

The most noteworthy finding from the questionnaire was that about
half of MESA participants stated that their grades had not improved as a

result of the program. Nevertheless, they continued to take advanced science and math courses. It is suggested that MESA activities and support led to increased confidence and perseverance in these difficult but important courses.

EUREKA! participant follow up analysis. PATRICIA B. CAMPBELL, CATHER-INE SHACKFORD, 1990. **Campbell-Kibler Associates.**

Statement of the problem/main purpose of the study:

This study reviews the long-term impact on plans to study math and science and sports participation among former and current participants in this intensive summer math/science and sports program.

Sample selected for the study:

The sample consisted of young women who had participated in the EUREKA! program beginning in 1987, 1988, or 1989 as rising eighth graders.

Overview of methods used in study/procedures:

Telephone interviews were conducted with 10 of 30 participants who participated in 1987; the remaining 20 could not be reached. Twelve 1988 participants who returned in 1989 completed questionnaires at the beginning and end of the 1989 program; sixteen 1988 participants who did not return were interviewed by telephone. Fourteen of thirty 1989 participants completed a questionnaire during a January 1990 "reunion."

Outcomes/major findings:

All of the 1987 participants located have taken math classes every year and plan to do so through high school. Although the 1987 program did not include science, nine of these ten young women plan to take at least four years of science. The young women who participated in both 1988 and 1989 EUREKA! consistently increased the number of mathematics courses they planned to take. The number of science courses they planned to take dropped slightly during the school year 1988–1989 but increased even more after participation in Science EUREKA! in summer 1989. The 16 nonreturning 1988 participants, only one of whom gave a program-related reason for not returning, planned to take nearly as many math and science courses as their peers who returned for a second summer. Math and science course-taking plans of the 1989 participants also increased.

There was some increased sports participation among participants from all three years, with two 1987 participants specifically crediting this interest to their EUREKA! experience.

Final report: First year evaluation of the Eureka! teen achievement program. BEATRIZ CHU CLEWELL, 1992. EUREKA!, Women's Center of Brooklyn College.

Statement of the problem/main purpose of the study:

This study tracks young women who participated in the first two years of the EUREKA! intensive mathematics and sports program to determine long-term impact on participation in mathematics, science, and sports.

Sample selected for the study:

The sample included all 1987–88 and 1988–89 participants who came to either of two participant reunions or could be contacted by phone. At the time of the interviews about one-third of each cohort consisted of rising seniors; the remainder of the older cohort had graduated high school, while the remainder of the younger cohort were rising juniors. National data were used for comparison purposes.

Overview of methods used in study/procedures:

Twenty of the 30 1987–1988 participants and 48 of the 73 1988–1989 participants participated in the study. Forty-two young women were interviewed in person, 26 by telephone, in relation to such program outcomes as attending a high school or program with math and science emphases, completion of four years of math and science courses in high school, enrollment in higher education, and participation in sports.

Outcomes/major findings:

Half or more of each cohort attended specialized math-science high schools or programs. EUREKA! participants on the average took more math than the members of the comparison groups. Figures for most science courses were similar, but EUREKA! participants were twice as likely to take physics as were members of the comparison groups. Most EUREKA! participants (92%) planned to attend a four-year college, with the remainder planning to attend a two-year college or vocational school. All but one of the high school graduates had been accepted to one or more colleges at the time of the interviews; this student was currently in the process of application.

At least three-quarters of each cohort stated they participated regularly in sports.

The African Primary Science Program: An evaluation and extended thoughts. ELEANOR DUCKWORTH, 1978. **North Dakota Study Group on Evaluation.**

Statement of the problem/main purpose of the study:

The African Primary Science Program was designed to help students develop an interest in, familiarity with, and competence engaging in science through intensive hands-on investigation of interesting materials. Students taught by teachers trained in Kenya's New Primary Approach or using the standard curriculum studied more material in less depth. Duckworth sought to determine how well children in both types of classes extended their learning to novel situations.

Sample selected for the study:

Students in three rural schools in Kenya whose teachers had received training in the African Primary Science Program constituted the experimental group. The comparison groups included students considered to be of equal or greater ability and whose schools were located in similar or more advantaged settings. Two schools, one of which included several experimental classes, were especially well matched, drawing both students and teachers from the same interrelated rural community.

Overview of methods used in study/procedures:

Twelve children were chosen at random from each of the 15 experimental and 13 comparison classes. In phase I, each group of classmates spent 35 minutes exploring a variety of previously unseen materials in any way they chose. A team of two observers noted each child's activities in turn. The activities were rated for complexity and diversity. In Phase II, the same children were presented with set tasks involving manipulatives. These tasks were scored based on how many aspects of the task the child completed successfully.

Outcomes/major findings:

In Phase I, experimental classes worked with the materials in significantly more complex and diverse ways than did comparison classes at the same grade level and, in all but one grade, at the next higher grade level. Classes from both well-matched schools whose teachers did not use the African Primary Science Program had virtually identical scores. In the more structured Phase II activities, children from experimental classes did as well as children from the comparison classes on all but one task and significantly better on several. The one experimental class which had used the African Primary Science Program for three years particularly stood out on these tasks.

An evaluation of the Hands On Science Outreach Program. IRENE F. GOODMAN, 1992. Sierra Research Associates. One component of *How are we doing?*, report submitted to the National Science Foundation under Grant #MDR 8954696.

Statement of the problem/main purpose of the study:

Goodman and her associates conducted a pilot study to evaluate participants' attitudes toward and understanding of this after-school, for-fee program and science in general. Other parts of this study include feedback from adult program coordinators and leaders and from previous participants' parents; these are not considered here.

Sample selected for the study:

Children representing 11 schools and the three program levels (grades K–1, 2–3, and 4–6) were selected in a quasi-random manner before the first class session. As noted below, substitute children were added at the fifth session.

Overview of methods used in study/procedures:

Participants were interviewed before the first, fifth and eighth (last) program sessions. If a child did not attend the fifth session, a classmate was interviewed instead. Thirty-four children completed all three interviews; an additional 16 completed one or two. No comparison group was interviewed.

Outcomes/major findings:

Participants' explanations of "What is science?" tended to become more sophisticated from the first to the last session. Younger participants saw Hands On Science as more fun than school science; older children described Hands On Science as involving more participation and experimentation than school science. Participants in all age groups agreed almost unanimously that they would recommend the program to a friend.

An evaluation of children's participation in the Hands On Science Outreach program. IRENE F. GOODMAN WITH KIM RYLANDER, 1993. Hands On Science Outreach, Inc.

Statement of the problem/main purpose of the study:

This study evaluates whether and how well children participating in the Hands On Science Outreach program are achieving program objectives. The objectives of this eight-week interactive science program include enjoying participation, understanding the many aspects of science, and greater ability to solve problems and to design and participate in science-related experiments and activities, all at an age-appropriate level.

Sample selected for the study:

The participant group included 74 children registered for Hands On Science classes in Fall 1992. These children represented four ethnically and geographically diverse cities and grades pre-kindergarten through 6. Thirty-nine students, all classmates of participants but themselves not currently registered for Hands On Science, served as a comparison group. Twenty-eight percent of the comparison group and 47 percent of the participation group had participated in previous sessions of Hands On Science.

Overview of methods used in study/procedures:

All 113 students were interviewed a week before the program started. Three participants were unavailable for follow-up; the remaining 110 students were interviewed again by the same interviewer about two weeks after the program ended. The pre- and post-interviews were nearly identical and included questions about what science is, attitudes toward science and who can do science, and scientific principles covered in the Fall 1992 session of Hands On Science. The students were also asked to do a simple science-related task with materials allowing a variety of approaches. Parents of nearly all 74 participants completed pre- and post-surveys on their children's participation. In addition, trained observers rated the participation of 51 of the students in their Hands On Science classes on the basis of observation and additional interviews.

Outcomes/major findings:

Current participants in Hands On Science made significant gains in their understanding of what science is and whether specific activities are science-related, in specific science content, and in their perceptions of who can do science; students in the comparison group did not. Virtually all participants enjoyed Hands On Science; their parents confirmed their interest as demonstrated by bringing materials home and talking about the activities. However, at the post-interview members of both groups tried fewer ways to accomplish the science-related task. Outcomes showed no difference by gender.

Science at age 15: A report on the findings of the age 15 APU science surveys. REED GAMBLE, ANGELA DAVEY, RICHARD GOTT, GEOFF WELFORD, **1985. Department of Education and Science, Assessment of Performance Unit.**

Statement of the problem/main purpose of the study:

This four-year study of 15-year-old students in Great Britain was designed to answer the following questions:

a. Can students apply science concepts in new contexts?
b. How well do students read, express and transfer information?
c. Can students read and use apparatus and measuring instruments correctly?
d. Can students make and interpret observations?
e. Can students interpret presented information?
f. Can students plan and perform investigations?

Sample selected for the study:

Participants were selected randomly from among students in a stratified sample of schools in England, Wales and Northern Ireland. Approximately two percent of all 15-year-old students in Great Britain took part in the study each year.

Overview of methods used in study/procedures:

Each student completed one of several test packages including pencil and paper and/or practical tasks related to one or more of the above goals. Thus students might be asked to read or use a measuring instrument, plan an experiment on paper or demonstrate how they would go about it, read or create graphs and charts, or answer questions based on reading passages, photographs, diagrams, or hands-on materials. The results for individual students and schools were aggregated to provide a broad view of participants' performance.

Outcomes/major findings:

Few students were able to apply concepts to novel situations. Most students had basic science skills, but had difficulty with any complications such as interpolating on a scale or manipulating two or more variables simultaneously. The results of pencil and paper measures and actually performing the tasks were very different. The authors note implications of these results for curriculum, teaching and assessment.

Science at age 11. Wynne Harlen, **1983. Department of Education and Science, Assessment of Performance Unit.**

Statement of the problem/main purpose of the study:

This two-year study of 11-year-old students in Great Britain was designed to answer the following questions:

> lp;&2qa.
> How well do students read, express and transfer information? b.
> Can students read and use apparatus and measuring instruments correctly? c.
> Can students make and interpret observations? d.
> Can students interpret and apply presented information? e.
> Can students plan and perform investigations?

Sample selected for the study:

Participants were selected from among students in schools in England, Wales and Northern Ireland.

Overview of methods used in study/procedures:

Each student completed one of several test packages including pencil and paper and/or practical tasks related to one or more of the above goals. Thus students might be asked to read or use simple measuring instruments, demonstrate how they would go about an experiment, read or create graphs and charts, or answer questions based on reading passages, photographs, diagrams, or hands-on materials. The results for individual students and schools were aggregated to provide a broad view of participants' performance.

Outcomes/major findings:

Most 11-year-olds could complete tasks requiring basic skills of observation, measurement, classification, and interpretation. About half could complete more complex tasks such as making predictions based on observation or information, suggesting controls for experiments, or applying science concepts to solve problems. Few could describe general patterns in observations and data, produce a plan for or carry out an adequately controlled investigation, or observe closely enough to note fine detail or exact sequence of events.

Effects of an intensive summer program on attitude toward mathematics of American Indian eighth grade students. JOHN J. HOOVER, CATHY ABEITA, 1992. **American Indian Science & Engineering Society.**

Statement of the problem/main purpose of the study:

AISES is piloting an intensive summer program in math and science for Native American students. They sought to determine if participation in this two-week program decreased anxiety, increased confidence, improved attitude, and heightened perceived usefulness related to mathematics.

Sample selected for the study:

The sample consisted of all 25 participants in the pilot program. All were rising eighth graders chosen on the basis of grade point average, interest, and recommendations. The 14 female and 11 male students represented 12 tribes and 11 states.

Overview of methods used in study/procedures:

Students completed four of the subscales of the Fennema-Sherman Mathematics Attitudes Scales (1986) before and after participation in the program. Students also rated program activities on five dimensions at the completion of the program.

Outcomes/major findings:

Increase in confidence and decrease in math anxiety were significant for all participants taken together. When examined by gender, decrease in math anxiety was significant for girls only. Differences on all other scales, for the group and by gender, were in the right direction but not significant. On a scale of 1–5 with 5 as the most positive rating, aggregate ratings of program areas ranged from 3.86 to 4.16. Participants will continue in this program for the next two summers, allowing evaluation of the long-term effects of the intensive summer experiences.

Assessing accelerated science for African-American and Hispanic students in elementary and junior high school. MICHAEL JOHNSON, **1991. Gerald Kulm and Shirley M. Malcom (Eds.),** *Science assessment in the service of reform* **(Washington, DC: American Association for the Advancement of Science), pp. 267–282.**

Statement of the problem/main purpose of the study:

This study examines the results and implications of using statewide examinations developed for high school students to assess and document the achievement of participants in the Science Skills Center, an accelerated science and mathematics program.

Sample selected for the study:

The main focus of the study was upon 16 students participating in an intensive biology program in the Science Skills Center during the school year 1988–1989. All were African-American or Hispanic; their ages ranged from 9 through 13 (4th through 7th grade). The study also includes some information on a similar group of students participating in the Science Skills Center's intensive programs during the following school year.

Overview of methods used in study/procedures:

The students were encouraged to view themselves as a "science team," working together to reach a challenging but attainable goal. They participated in a year-long intensive study of biology, using the same textbook and review book and participating in the same laboratory sessions as do high school students studying for the Regents examination. These young students also practiced test-taking skills.

Outcomes/major findings:

In May 1989, all 16 students took and passed the tenth-grade New York State Regents biology examination, receiving the appropriate recognition and academic standing. In June 1990, after a year of intensive study a similar group of students took the Regents biology examination, the New York State Sequential I mathematics examination (covering algebra, geometry, logic, and problem solving), or both. All passed the examinations. High school students' rate of success on these examinations is not noted.

Issues raised by this study included the self-fulfilling prophecy of low expectations for school science achievement among minority students. The author also noted gender issues in recruiting and teaching in the Science Skills Center and the value of positive peer pressure.

Gender analysis of MATHSTART: Girls and boys make equal gains in HOSO preschool mathematics program that stresses hands on, teacher preparation and parental involvement. PHYLLIS KATZ, 1992. **Hands On Science Outreach, Inc.**

Statement of the problem/main purpose of the study:

The hypothesis presented is that materials such as MATHSTART help to prepare preschool-age children for future mathematics achievement. The evaluator also looked for any evidence of gender difference in completing MATHSTART tasks.

Sample selected for the study:

Ten Head Start classes participated in the development of MATH-START. Five "treatment" classes did MATHSTART activities; five other classes in the same centers served as a control group.

Overview of methods used in study/procedures:

Approximately 100 children were randomly selected from the treatment group and a similar number from the control group in the fall to provide baseline data. In the spring, random selection was again made of approximately 100 children from each group, so that the study design evaluated the effects of the program as a whole and not on individual participants. All selected children were asked to identify, point out, and make simple geometrical shapes, topics that were covered in the fall session of MATHSTART; each task was scored for successful, partial, or unsuccessful completion.

Outcomes/major findings:

It is difficult to interpret Katz's findings because the treatment and control groups seem to have differed in some systematic way at the baseline. In the fall, before MATHSTART, the control group completed the tasks much more successfully. The treatment group improved its scores much more than did the control group (all differences in gain were highly significant) so that the average spring scores for both genders and groups were virtually identical.

The only significant differences by gender involved description of a figure and two solids and favored girls in both groups in the spring. Katz notes that these tasks may be measuring preschool girls' generally greater verbal ability rather than their comprehension of geometry.

Draw-A-Scientist test: Future implications. CHERYL L. MASON, JANE
BUTLER KAHLE, APRIL L. GARDNER, 1991. *School Science and Mathematics, 91*(5), 193–198.

Statement of the problem/main purpose of the study:

This study tested an intervention program designed to decrease stereotyping of scientists as male and sinister ("mad scientist") or eccentric ("nerd") and to increase perception of science and science-related careers as attractive options for well-adjusted individuals of both genders.

Sample selected for the study:

The study sample included students in classes taught by 14 high school biology teachers. These teachers were selected randomly from a group of volunteers.

Overview of methods used in study/procedures:

Seven teachers were chosen for the experimental group. These teachers received training and encouragement in presenting the study objectives in their classes. The other seven teachers received no special training. Demographics of teachers, cohorts of their students, and school systems did not differ significantly by group.

After a year of studying biology under these 14 teachers, all students were asked to draw their "image of a scientist" on a blank sheet of paper. Randomly selected students in the experimental and control groups were also interviewed on their attitudes toward science and their image of scientists.

Outcomes/major findings:

All students and female students in the experimental group were significantly more likely than all students and female students in the control group to draw female scientists. No male student in the control group drew a female scientist, though group by gender interaction was not significant for males. Students in the experimental group drew significantly more "neutral" (not sinister or eccentric) scientists than did students in the control group. Some students in the experimental group depicted scientists working in defined places other than laboratories.

Interviews suggested that cartoon and movie images help to shape the common perception of scientists as unbalanced men. Students saw much classroom science as boring and complicated but stated that they enjoyed hands-on activities and appreciated teachers who explained science in everyday language and did not ask questions requiring one "right" answer.

A twelve-year longitudinal study of science concept learning. JOSEPH D. NOVAK, DISMAS MUSONDA, 1991. *American Educational Research Journal,* **28**, 117–153.

Statement of the problem / main purpose of the study:

The study is designed to show the impact of instruction in basic science concepts on young students over a period of years.

Sample selected for the study:

The sample began with 239 students in four Ithaca schools, all of whom were in the first grade in 1971–72. Only 55 students participated in the last round of interviews.

Overview of methods used in study / procedures:

One hundred ninety-one students used audiotutorial lessons in basic science concepts, consisting of hands-on materials and accompanying taped instructions and information, in their first- and/or second-grade classrooms. Forty-eight of their peers did not use the lessons in either grade. Teachers of both groups provided little or no additional science education. As many as possible of the students were interviewed regarding concepts included in the lessons in grades 2, 7, 10, and 11 or 12, though attrition, availability of funding, and refusal by a few participants in the older grades steadily decreased the sample. Interviewers translated the understandings expressed into concept maps, which they then rated. SAT scores of the remaining sample were compared between groups and to the mean scores of all 1983–84 seniors and to the Ithaca high school class of 1984.

Outcomes / major findings:

Over the course of the study, students who had used the audiotutorial lessons showed more understanding and less misunderstanding of science concepts and were able to connect these concepts better than were their peers who studied little science in the primary grades. Evidence that boys of high school age had more accurate and richer understanding of concepts than did their female peers deserves further study.

1982–83 through 1986–87 results on the Comprehensive Test of Basic Skills (CTBS test). PROJECT INTERFACE, 1987. Project Interface, J. Alfred Smith Fellowship Hall, 8500 "A" Street, Oakland, CA 94621.

Statement of the problem/main purpose of the study:

This study uses results from the Comprehensive Test of Basic Skills (CTBS) to document the academic gains attributable to participation in this math and science enrichment program by students in grades 7–9.

Sample selected for the study:

All students participating in Project Interface in each given year constituted the experimental group. All students in the same grade levels at "home sites" (the schools attended by the majority of Project Interface participants) and the school district as a whole served as comparison groups.

Overview of methods used in study/procedures:

Gain in median scores, "grade equivalents" (progress measured in months), and percentile ranks on the CTBS for Project Interface students were compared with those of students in the same grade at the "home site" schools and in the district as a whole for that year.

Outcomes/major findings:

In all program years, gain in scores for Project Interface participants was at worst comparable to and often considerably better than those of the comparison groups. In all years Project Interface participants surpassed the ten-month expected "growth"; in no year did more than one home site and the district as a whole also achieve this goal.

In the first two years of the project, participants far surpassed the change in rank of three percentiles considered significant; gains in the other two measures in these years were also far higher than in subsequent years. The third year, 1984–85, saw a district-wide teachers' strike during which all home sites and the district as a whole either made minimal progress or lost ground in all three measures. In this "devastating" year seventh graders participating in Project Interface suffered the most, only gaining six months' growth and falling nine percentiles; eighth and ninth graders managed to achieve ten and seventeen months' growth and to move up three and six percentiles respectively.

Class placement of students for 1986–87 and 1987–88. PROJECT INTER-FACE, 1987. Project Interface, J. Alfred Smith Fellowship Hall, 8500 "A" Street, Oakland, CA 94621.

Statement of the problem/main purpose of the study:

This study tracks the enrollment of Project Interface participants and graduates in college preparatory mathematics classes.

Sample selected for the study:

The study includes all seventh through ninth-grade students completing the previous term of Project Interface; that is, students were in grades 8–10 in the year of the study. The study does not include a comparison group.

Overview of methods used in study/procedures:

Participants' mathematics course enrollment for the year following Project Interface participation was categorized as "not enrolled in college-prep," "enrolled in college-prep" (taking a mathematics course that could have been taken in an earlier grade) and "on-track" (right course, right year).

Outcomes/major findings:

In 1986–87, 76 percent of participants were taking college preparatory math; of these 60 percent were on (or, in one case, ahead of) track. The projected enrollments for 1987–88 were 71 percent in college preparatory math, 80 percent of these on track. The percentage of students on track by grade decreased by age in 1986–87 and for eighth and ninth graders in 1987–88; tenth graders in 1987–1988 were nearly as likely to be on track as eighth graders.

9th grade graduate status summary: 1982–83 through 1985–86. PROJECT INTERFACE, 1986. Project Interface, J. Alfred Smith Fellowship Hall, 8500 "A" Street, Oakland, CA 94621.

Statement of the problem/main purpose of the study:

This study tracked the continuing progress in college preparatory mathematics of graduates of Project Interface (students who had participated in this math and science enrichment program through grade 9).

Sample selected for the study:

The study sample included all 63 program graduates, of whom 12 could not be contacted (9 of these from the 1982–83 cohort of 20).

Overview of methods used in study/procedures:

Program graduates were interviewed by telephone and asked what mathematics course they were presently taking, their current grade in that course, and whether they were participating in another supplementary program.

Outcomes/major findings:

Of the 51 program graduates who could be contacted, 43 (85%) were enrolled in college preparatory math classes. The more years had elapsed since program graduation, the more likely the program graduate contacted was to be taking a college preparatory math class. Sixty percent of program graduates from the first three cohorts were earning a "C" or better in math; 40 percent were earning a "B" or better.

Evaluation of Year One implementation of Operation SMART in rural communities by the Girls Club of Rapid City, Inc. STEVEN ROBERT J. ROGG, 1991. Girls Incorporated of Rapid City.

Statement of the problem/main purpose of the study:

Formative and summative evaluation were used to determine whether the program was meeting stated objectives, most notably changes in attitude toward participation in science and changes in perceptions about science, scientists, and gender roles.

Sample selected for the study:

The study universe consisted of the 356 members of Girls Incorporated of Rapid City who participated in at least one session of Operation SMART programming. Subsamples of this group were surveyed in different aspects of the study.

Overview of methods used in study/procedures:

For all study participants, number of sessions attended was graphed against performance on math and science attitudes surveys based on the Fennema-Sherman Mathematics Attitudes Scales. Some participants also completed the Draw a Scientist Test and Sex Role Survey from the Operation SMART Research Tool Kit and a science experiences survey based on one by Kahle (1983).

Outcomes/major findings:

Number of sessions attended showed a positive but weak correlation with performance on the math and science attitude surveys and a somewhat stronger correlation with the perception that a variety of science-related or prestigious jobs can be held by women as well as men.

More SMART participants than "would be expected" drew female scientists, but many drawings stereotyped scientists as "weird." Recommendations for future evaluation include a comparison group of members of Girls Incorporated of Rapid City not participating in Operation SMART.

The TERC Environmental Network project pilot summer program: Final evaluation report. WILLIAM SPITZER, JUNE FOSTER, 1991. TERC.

Statement of the problem/main purpose of the study:

This study examines student outcomes and the role and impact of technology in a summer program on air pollution designed for seventh and eighth graders and the feasibility of implementing this program on a larger scale.

Sample selected for the study:

Participants for the pilot program were recruited by a science museum, a science center, and a nature center, all in Connecticut. The length of program participation ranged from 15 to 22 hours.

Overview of methods used in study/procedures:

All 28 participants, ranging from fifth to ninth grade, participated in the study. Sources of data used to assess individual students' learning and the program as a whole included student work samples, classroom observation, videotaped interviews with students and teachers, student pre- and post-questionnaires, and teacher questionnaires.

Outcomes/major findings:

All students gained some conceptual understanding and data analysis skills. The depth of understanding and analysis was to some extent correlated with grade level. Students and teachers enjoyed using the computer software adapted for the program and appreciated its possibilities for sharing of data and consultation with other students and working scientists. Recommendations for further implementation include a longer program with more hands-on activities, availability of more computers, simulations and games related to the topic, and strategies for recruiting more participants of the target ages. It was also suggested that the program model be extended to other science topics.

EQUALS at Cleveland State University: 1985–6 evaluation report. ROSEMARY E. SUTTON, ELYSE S. FLEMING, 1987. College of Education, Cleveland State University.

Statement of the problem/main purpose of the study:

This study compares the problem-solving skills and attitudes toward mathematics of students whose teachers have and have not received

EQUALS training. Special attention is paid to any differential impact by gender or race.

Sample selected for the study:

Study participants consisted of Cleveland-area students in grades 4–12, some taught by EQUALS-trained teachers and some by other teachers at the same school and grade level.

Overview of methods used in study/procedures:

All study participants took problem-solving tests and completed several math attitudes scales before EQUALS activities were implemented in the treatment group's classes and at the end of the year. The problem-solving tests were designed for grades 4–6, 7–9, and 10–12 and came in two different forms so that participants solved similar but not identical problems at the pretest and post-test.

Outcomes/major findings:

The problem-solving skills of both treatment and comparison classes in grades 4–6 and 10–12 increased significantly over the year. Students of EQUALS teachers in grades 7–9 increased their problem solving skills significantly, while the results of their peers in the comparison group showed a decrease in skills. White females and African-American males in classes taught by EQUALS teachers showed the highest increases in problem-solving scores.

Students in grades 4–6 taught by EQUALS teachers became less likely to see math as a male domain; their peers became more likely to do so. The beliefs of both groups of this age in the intrinsic value of math declined over the year, but less so for students of EQUALS teachers. These beliefs increased over the year for students of EQUALS teachers in grades 10–12 but decreased for their classmates.

Evaluating an intervention program: Results from two years with EQUALS. ROSEMARY E. SUTTON, ELYSE S. FLEMING, **1989. College of Education, Cleveland State University.**

Statement of the problem/main purpose of the study:

This study looks at changes over a year in the problem-solving skills and attitudes toward mathematics of 4th–9th grade students in the Cleveland State University study described above.

Sample selected for the study:

Study participants consisted of Cleveland-area students in grades 4–10, some taught by EQUALS-trained teachers and some by other teachers at the same school and grade level.

Sample selected for the study:

All study participants took problem-solving tests and completed several math attitudes scales before EQUALS activities were implemented in the treatment group's classes and at the end of the year. The problem-solving tests were designed for grades 4–6 and 7–9 and came in two different forms so that participants solved similar but not identical problems at the pretest and post-test. Data from each of the two years of the study were analyzed separately.

Outcomes/major findings:

The problem-solving skills of both treatment and comparison classes in grades 4–6 increased significantly over both years. In both years, students of EQUALS teachers in grades 7–9 increased their problem solving skills significantly, while the results of their peers in the comparison group showed a decrease in skills. In the first program year, white females and African-American males in classes taught by EQUALS teachers showed the highest increases in problem-solving scores. Insufficient numbers precluded analysis by both gender and race in the second year, where female students taught by EQUALS teachers had the highest increase in problem-solving scores.

In the first year only, students in grades 4–6 taught by EQUALS teachers became less likely to see math as a male domain; their peers became more likely to do so. Also in the first year, the beliefs of both groups of this age in the intrinsic value of math declined over the year, but less so for students of EQUALS teachers. There were no significant changes in attitude over the year for fourth through sixth graders in the second year of the study or for seventh through ninth graders in either year.

The evaluation of Project SEED, 1990–91. WILLIAM J. WEBSTER, RUSSELL A. CHADBOURN, 1992. Department of Evaluation and Planning Services, Dallas Independent School District.

Statement of the problem/main purpose of the study:

This study, designed to replicate earlier studies of this mathematics enrichment program, asked the following evaluation questions:

    a. What impact does one, two or three semesters of SEED instruction have on mathematics achievement, and is this impact cumulative?
    b. Do SEED participants enroll in more higher level math classes than peers who did not participate?
    c. What is the long-term impact of participation in Project SEED?

Sample selected for the study:

The first set of experimental groups included students who had just completed one, two or three semesters of participation in Project SEED (grades 4, 4–5, or 4–6). A second set of experimental groups included students who had completed three semesters of participation from one to four years previous. Students matched to each Project SEED participant by gender, ethnicity, grade, socioeconomic status, and math achievement level before Project SEED participation formed comparison groups for each experimental group. All students in both groups were African-American or Hispanic; most were from low-income families.

Overview of methods used in study/procedures:

Achievement on the Iowa Test of Basic Skills (ITBS) was compared after the current semester of Project SEED participation for the first set of experimental groups and the matched comparison groups. For the second set of experimental groups and their matched comparison groups, achievement was measured by the Tests of Achievement and Proficiency (TAP) one, two, three or four years after the experimental group completed three semesters of Project SEED.

Outcomes/major findings:

Participants in one semester of Project SEED scored on average three to four months higher in math achievement on the ITBS than did nonparticipating peers. The advantage in math achievement for two- and three-semester participants over their peers was even greater. Four years after participating in three semesters of Project SEED, participants still outperformed their peers on the TAP by five months, which difference was significant. Students who had participated in Project SEED also enrolled in more higher level math courses in middle school and high school than did their peers.

"Families in FAMILY MATH" research project: Final report. KATHRYN SLOANE WEISBAUM, 1990. Lawrence Hall of Science, University of California, Berkeley.

Statement of the problem/main purpose of the study:

This study examined the effect of participation in FAMILY MATH programs on parents' and children's attitudes toward mathematics.

Sample selected for the study:

Interview subjects included 84 parents of children ages 7–10 representing four different ethnic groups (European-American, African-American,

Mexican-American, and Native American) and program sites. These parents did not differ from other parents in the same FAMILY MATH classes in their responses on a math attitudes survey based on the Fennema-Sherman Mathematics Attitudes Scales.

Overview of methods used in study/procedures:

Most parents were interviewed about their attitudes toward mathematics and related topics before and at the end of their FAMILY MATH participation and four months later. The 15 Native American parents were interviewed once, several months after their FAMILY MATH participation.

Outcomes/major findings:

Common changes in parents' attitudes toward math included the realizations that math can be fun, the math used in everyday activities is legitimate mathematics, there is more than one way to teach math, and math problems can be solved cooperatively and in more than one way. As a result, parents felt more secure in their own ability to do math and to help their children learn math. The latter was particularly true of the Native American and Mexican-American parents, who had often found it difficult to help their children with school math because of language and cultural barriers.

Effects of using programmed cards on learning in a museum environment. R.J. DeWaard, N. Jagmin, S.A. Maisto, & P.A. McNamara, 1974. The Journal of Educational Research, 67(10), 457–460.

Statement of the problem/main purpose of the study:

Researchers assessed the educational effectiveness of providing visitors with a series of cards which offer information and ask visitors to answer questions about an exhibit. Some of the card systems which were tested contained more information than others; and some provided immediate feedback regarding correct or incorrect answers, while others provided no feedback.

Sample selected for the study:

120 visitors to the *Age of Man* exhibit at the Milwaukee Public Museum, ages 13 and older. Median age of participants was 16.5 years.

Overview of methods used in study/procedures:

Visitors were selected as they entered the area of *Age of Man*, an exhibit which illustrated evolutionary concepts by presenting a series of skulls, and describing the differences between them. Visitors who volunteered to participate were informed that an "expert medal" would be given to those who performed especially well, and then were randomly assigned to one of six groups: High Information, Low Information, Feedback, No Feedback, and the control groups Posttest Only, and Exhibit Only.

Except for those in the Control group, all participants were given an 18-frame linear program in the form of index cards. High Information cards presented important concepts from the exhibit, as well as questions regarding those concepts. Low Information cards presented only the questions. Feedback cards were designed to immediately inform participants of the accuracy of their answers, while No Feedback cards provided no such information. Subjects in the Exhibit Only control group viewed the skulls without the linear program, and those in the Posttest Only group did not view the exhibit. Participants in all six groups were then tested on their knowledge of the exhibit, with twelve multiple-choice questions. Each individual who answered at least eleven questions correctly was given an expert medal.

Outcomes/major findings:

Scores were analyzed using a 2×2 complete factorial design. Among participants who had used the linear program cards, 30 percent earned an expert medal. Of those who had not viewed the exhibit, or had viewed the exhibit without the aid of the cards, none received a medal. However, no significant differences were found among those who had, or had not, received feedback, or between those who were given high or low amounts of information. No interaction effects were found.

Researchers concluded that both the questions and the information presented in the cards produced greater learning than did viewing the exhibit without the cards. The similarly low scores of those in both control groups suggested that visitors without additional aids did not attend to important aspects of the exhibit.

It was suggested that the lack of differentiation between the four experimental groups was due to the success of the card system in directing the attention of all visitors to the exhibit. It was also hypothesized that the simplicity of the questions provided on the cards allowed respondents to easily determine the correctness of their responses by looking at the exhibit, thereby limiting the extra benefits of providing explicit feedback. It was also thought that results might have been affected in some way by the relatively young age of those who agreed to participate in the study.

Self-testing Raphaël: How a computer stimulates visitors in an art exhibition. H. Gottesdiener & J. Boyer, 1992. *ILVS Review*, 2(2), 165–180.

Statement of the problem/main purpose of the study:

Despite an increase in the use of computers within museums, very little is known about the benefits and drawbacks of their application as didactic tools in art museums. This study assessed the impact of a self-test computer game which served as an advance organizer for a gallery of paintings, in terms of visitor behavior and learning.

Sample selected for the study:

Visitors were observed throughout the gallery during a four day period. A sample of players and nonplayers were given a questionnaire.

Overview of methods used in study/procedures:

The project was carried out at an exhibition entitled "Raphaël and French Art," at the Grand Palais in Paris. This exhibition was intended to illustrate the influence of the painter Raphaël on other artists. The museum developed a game consisting of a computer which controlled four sets of slides, and presented players with a series of questions about those slides.

The game, which was placed in a gallery, required visitors to complete four sets of tasks, in conjunction with a sequence of 19 questions. Tasks included identifying which Raphaël painting had inspired a particular artist, recognizing which of a series of paintings was inspired by Raphaël, identifying Raphaël's work, and distinguishing between originals and reproductions of his work.

The computer automatically recorded the task performance of players, who were also observed unobtrusively to record their overt reactions. Visitors were tracked throughout the remainder of the gallery as well, to record the path that both game players and nonplayers followed during their visit. A sample of visitors was also interviewed as they left the gallery, to record their reactions to the game as well as their knowledge of the paintings they had seen.

Outcomes/major findings:

Observation revealed that the game was very popular among visitors, with 40% of the players completing the entire game. Over 70% of those who played the game pressed buttons which offered additional information about the correct answer. Often this information was selected even by those who had answered the questions successfully.

Through exit interviews, it was discovered that players and nonplayers were similar in terms of nationality and gender. However, players who completed the entire game tended to be younger than nonplayers. It was hypothesized that older players were more likely to quit after failing at a task, and that labelling the tasks as a "game" may have discouraged older visitors. Observation of players showed that about half of the games were played by small groups of visitors as a team, with frequent verbal exchanges between different players, and between players and spectators.

As expected, exit interviews also showed that game players were better able to recall information about Raphaël's paintings than were nonplayers. However, little difference was found between their scores on questions which had to do with paintings by artists other than Raphaël. Tracking of visitors found that those who had played the game were also much more likely to stop and view Raphaël's paintings than were those who had not. Significantly, this increase in viewing by players did not detract from their viewing of other paintings, as players were equally as likely to stop and view non-Raphaël paintings as were nonplayers. It was suggested that this study had shown that, while not all visitors choose to use such technology in an art museum, computer games can provide a higher quality of visitor experience for many.

Behavior of the average visitor in the Peabody Museum of Natural History, Yale University. MILDRED PORTER, 1938. AAM Monograph New Series No. 16. Washington, DC: American Association of Museums.

Statement of the problem/main purpose of the study:

The galleries of the Peabody Museum were arranged so that visitors could circulate through the museum in a sequence consistent with evolutionary history beginning with the Invertebrates Hall, the Great Hall (including dinosaurs), two mammal halls, and the final hall on humans. One purpose of the study was to assess the impact of this arrangement on visitor behavior. Would "museum-fatigue" be reduced if visitors followed this sequence?

Another purpose of this study was to evaluate the impact of museum guides on visitor behavior. Would such guides increase attention to the exhibits? If so, would this increased attention be limited to those objects mentioned in the guide?

Sample selected for the study:

Over a two-year period, Sunday afternoon visitors were tracked throughout the Museum.

Overview of methods used in study/procedures:

Two museum guides were prepared. One type of guide was designed to emphasize the planned sequence of exhibits following the story of organic evolution. The other guide was unsequenced, merely giving information without relating it to the story of evolution. A third condition (no museum guides) was alternated every third Sunday with the two guide conditions.

Guides, in the form of short leaflets, were distributed by a person to visitors as they entered the museum. Visitors were tracked through their visit and times in galleries and attention to exhibits and labels were noted.

Outcomes/major findings:

With no museum guide, visitors averaged 21.4 minutes on the first floor of the Peabody Museum. With the sequence guide, visitors averaged 32.5 minutes and with the nonsequenced guide visitors averaged 37.9 minutes. There was no statistically significant difference between the two guides, but either guide produced significantly longer visitation than no guide. Guided visitors engaged in more label reading than unguided visitors, although this difference tended to decrease with time in the museum. Guided visitors also examined more cases (both those mentioned and not mentioned in the guide) than unguided visitors.

What makes museum labels legible? L.F. WOLF & J.K. SMITH, 1993. *Curator,* 36/2, 95–110.

Statement of the problem/main purpose of the study:

A study was undertaken by the Metropolitan Museum of Art in an attempt to provide empirical support for a new guide to producing legible labels. This study focused on variables including type face, type size, contrast between text and background, spacing between lines, lighting, and height and angle of installation. These factors were tested with individuals who were elderly or had poor vision, as well as with members of the Museum's curatorial staff.

Sample selected for the study:

One hundred twenty-one volunteers. Each of these individuals was assigned to one of four different groups: (1) low vision (those who had corrected vision from 20/40 to 20/200), (2) elderly (age 65 or over), (3) low vision elderly, and (4) curatorial staff members. It was immediately discovered that all of these groups provided quite similar ratings of the test labels, and as a result all volunteers were combined into a single group for analysis.

Overview of methods used in study/procedures:

Fifty labels were developed to accompany a painting by Claude Monet, each with identical copy, but differing in terms of a number of research variables. These variables included five different type faces, five different type sizes, wall labels of four different heights, and floor and case labels in either an angled, or a flat, horizontal placement. Manipulations also included two different lighting levels, two levels of text/background contrast, and two different amounts of spacing between lines of text. Subjects were asked to complete a questionnaire which solicited demographic information, and inquired about their art background, frequency of museum visits, and use of museum labels. The importance of all of these factors was assessed in four different studies: an interaction study, and three focused studies.

For each study, subjects were asked to rate labels on a 10-point scale, according to four criteria: (1) legibility, (2) easy or tiring to read, (3) aesthetic appeal, and (4) overall rating. Since it was found that ratings were comparable on all four scales, only legibility ratings were used to conduct analyses.

Outcomes/major findings:

Type size was the most important factor in predicting ratings of legibility. While interaction effects between size and the other variables were noted, as a rule, increases in size significantly correlated with dramatic increases in ratings. Although interaction effects made it difficult to determine a single optimal type size, it was suggested that when all other variables were at an optimal level, labels of 18 point type were rated as being quite legible.

Manipulation of type face produced somewhat less distinct results. Only Centaur, which has rather small lower-case letters, received poor ratings. Helvetica scored well, although serif fonts were found to be adequate, as well.

Analysis of the contrast between text and background showed that higher contrast labels received higher ratings. Somewhat surprisingly, ratings of high contrast light text on dark background received ratings which were almost identical to those given to dark text on light background.

When considered across all the other type characteristics, the amount of spacing between lines appeared not to produce significantly different ratings of legibility. However, closer analysis revealed that when labels were placed at eye level, standard spacing was thought to be more legible. On the other hand, labels which were placed above eye level, broad spacing was rated significantly higher.

When participants were asked to rate wall labels at four different heights, their ratings suggested that labels at eye level were greatly preferred over any other placement. However, since labels placed over 72″ from the floor received much lower ratings of legibility, it was decided that, considering the amount of variation in museum visitor height, lower labels are a better choice. For case labels, those which were placed at a 45 degree angle were clearly preferred. For floor labels, those which lay flat were preferred.

Variations in lighting did not produce significant differences in ratings of legibility. It was suggested that these results were confounded by problems resulting from glare, and from the amount of time participants were given to adjust to the lighting level.

Interesting results from the initial questionnaire included the discovery that respondents claimed that they usually read museum labels, that they felt labels were very important, and that they often look back and forth between a label and the work of art to which it refers.

Assessment of visual recall and recognition learning in a museum environment. W.A. BARNARD, R.J. LOOMIS & H.A. CROSS, 1980. *Bulletin of the Psychonomic Society,* **16(4),** 311–313.

Statement of the problem/main purpose of the study:

This study follows up on some early studies done in museums by Melton and Robinson of visitors time and attention to exhibits as an indicator of interest and potential learning. Recall and recognition memory had been tested in the laboratory but not in a museum that would represent the real context of sequentially experienced stimuli. Two conditions were investigated: total and limited exposure to the environment of a small history museum.

Sample selected for the study:

Sixty undergraduate students participated in the study. Half of them viewed the exhibits with limited exposure (LE) by following a standardized route past 8 of the 37 exhibits. The other half saw all 37 exhibits, or total exposure (TE). Thirty visitors at the museum were unobtrusively observed at five randomly selected exhibits to compare their time with that of the participants in the study.

Overview of methods used in the study/procedures:

Subjects moved at their own pace, recorded the time spent at each exhibit on a schematic diagram of the museum, and estimated the percent of familiar items in each exhibit case. All participants were tested immediately after looking at the exhibits, by listing objects (free recall), and differentiating photographs of objects (ones that were, or were not, in the exhibit).

Outcomes/major findings:

As expected, recall performance was lower in the TE group. Recognition was also lower with this group (an unexpected finding). Attention time was positively correlated with recall for the LE and the TE. Recognition performance was not significantly correlated with time for TE or LE. Differences in the findings from expected was attributed to the influence of context of being in the museum.

Consistent with earlier museum studies, these findings suggest that attention time observations can be considered a limited index of visual learning as evidenced by recall. A comparison of the subject's time with visitors who were not part of the study suggest that visitors in general have shorter attention times.

A study of children's interests and comprehension at a science museum. J.A.M. BROOKS & P.E. VERNON, 1956. British Journal of Psychology, 47, 175–182.

Statement of the problem/main purpose of the study:

The Children's Gallery at the Science Museum in London was investigated for the kinds and numbers of visitors at different times of the year; their reasons for visiting; the behavior of the children in the gallery, and their preferences for and comprehension of certain exhibits.

Sample selected for the study:

General visitors to the gallery were counted for attendance figures. Unaccompanied children were observed and interviewed.

Overview of methods used in the study/procedures:

The number of persons entering the gallery were counted during different time periods and grouped by age and sex and compared for different patterns during hours of the day and season of the year.

Observations were made of 50 children (35 boys and 15 girls) as they used the 27 different exhibits. The amount of time they spent was recorded. These observations yielded average time, range of time, and popularity of exhibit elements.

One hundred unaccompanied children (68 boys and 32 girls), picked at random, were "buttonholed" to participate in an oral questionnaire. They were surveyed for their age, schooling, residence, visiting habits, and their ability to correctly comprehend the meaning of five selected exhibits.

Outcomes/major findings:

Overall attendance is about 229,000 children per year, with enormous variations on different days. The number of adult visitors to the gallery normally exceed children of all ages. Possibly, some adults are attracted to the gallery because it is a good match for their level of appreciation and understanding of science.

Less than half of the exhibits were looked at by 50% or more of the children, and less than half are studied for more than 1 minute, although there is a wide range of variation. Working models held their attention the longest. Dioramas showing "daily life interests" were popular. The average time in this "large room" was 22 minutes for boys and 15 minutes for girls with 40% as many girls as boys visiting at all.

More than half of the children were coming from distant parts of London to reach the museum, spending noteworthy trouble and expense to reach the Kensington area. Boys indulged in repeat visits more than girls, coming for reasons of school work, amusement, and scientific interest.

Of the five exhibits reviewed by the children, two were fairly comprehensible to those who attempted them. Pulleys were often used but very few children had a grasp of how they work, probably because the principle is too difficult (and the labels were too complex to explain them). Some resemblance was noted between comprehension and votes for favorable exhibits, "suggesting that children do not particularly like exhibits which they do not understand, merely because they can play about them actively."

Orientation in a museum: An experimental visitor study. M.S. COHEN, G.H. WINKEL, R. OLSEN, & F. WHEELER, 1977. *Curator*, 20(2), 85–97.

Statement of the problem/main purpose of the study:

Good orientation is an essential part of a successful museum visit. This study assessed the effectiveness of different orientation aids, including maps, signs, directories and information people at the National Museum of History and Technology in the Physical Science wing.

Sample selected for the study:

Randomly selected visitors to the museum in the summer and early winter. No actual subject counts are reported, but "21,000 pieces of data" were gathered.

Overview of methods used in the study/procedures:

Maps were put at the entrance to the museum and the entrance to the Physical Science wing, detailing all the museum's halls, with color coding and color photographs of all the exhibit halls. Maps were placed in the visitors path; and different designs were tried. Signs were hung from the ceiling and directional arrows were used. Uniformed people were stationed at several places in the wing. Visitors were interviewed as they entered and left the wing during a baseline condition (directory and information desk), while maps and/or signs were in place.

Outcomes/major findings:

Signs were the most influential orientation aids, although the combination of maps and signs was slightly better for improving orientation by telling visitors what to expect, allowing them to make choices and organize their visit, eliminating backtracking, and reduce disorientation in the complex halls. More visitors used maps than either the information desk or the directory.

An orientation system with integrated elements of maps and signs was recommended to help satisfy visitors "insatiable demand" for orientation because they will "feel more secure if there is redundancy in the information system."

The motivation and education of the general public through museum experiences. L. NEDZEL, 1952. Unpublished doctoral thesis. University of Chicago, Chicago, Illinois.

Statement of the problem/main purpose of the study:

An experimental exhibit, "The Magnetism Room," was created at the Museum of Science and Industry in Chicago under a cooperative agreement between the museum and the University of Chicago. Nedzel was Associate Curator of Physics and Research Associate in Education at the museum, and the exhibit was to provide a setting for research in behalf of improving the educational effectiveness of exhibits for the adult public. A rigorous evaluation was carried out to assess the actual efficacy of the exhibit techniques used, and the methodologies and finding reported in this 105-page dissertation.

The exhibit space was a 300-square foot room with exhibit with exhibit cases on the east and west walls and an island unit in the middle of the room, with entrance/exit doorways at the north and south sides of the room.

Sample selected for the study:

Casual, self-selected visitors to the Magnetism Room were observed for behaviors in the room; 486 weekend only visitors were tested for knowledge, self-selecting to participate by responding to a sign. College students were used to test the questionnaire. In the controlled experiments, 291 visitors were actively recruited by the researcher who attempted to be unbiased, approaching all types of people.

Overview of methods used in the study/procedures:

Time-lapse photos, at 5-, 7½- and 15-second intervals, were used to record visitor movement in the room, and the photographs were analyzed and transcribed onto a chart that showed visitors' paths and time spent.

A 16-item, multiple choice, paper and pencil test was administered to visitors who had self-selected or had been recruited to participate the evaluation. Items were scored individually and together, noting overall percentage correct answers and differences in visitors performance on each item under each experimental condition.

Research was done partly under the "open" condition while the room was freely accessible to casual, self-selected visitors (time-lapse and questionnaire studies using mainly the existing exhibits). The experimental conditions were: no motivation, pre-test motivation, and sign motivation. During the "closed" condition, the room was not open to the public, and visitors were recruited outside the room and brought into it to use the exhibits under different experimental conditions: controlled length of stay at each exhibit—15-seconds, 40-seconds, and saturation; controlled sequence of viewing; and, with a lecture given before viewing.

Outcomes/major findings:

Median time spent per exhibit by casual visitors who stopped at least once in the Magnetism Room for the majority of exhibits was 15–20 seconds. Most visitors used the room as a passage, taking 5 seconds to pass through. Visitors who did stop, mainly looked at exhibits along only one wall (east or west). About 27% used exhibits on both walls. The hoped-for paths or sequences of viewing to be influenced by the architectural plan did not occur.

Students who had studied magnetism in a survey course achieved scores ranging from 54% to 100% on the 16-item test (although two questions were dropped, due to consistently low scores, and the remaining items were not of equal difficulty). Students who had not studied magnetism scored from 20% to 82%, averaging about 57%.

The sample audience who participated in the "open room" tests were 24% students, 28% white collar and professionals. Forty-five percent replied "yes" to having studied magnetism. Under the conditions of post-test only, pre- and post-test, and a motivating sign, the learning gains were small, never exceeding 57%, and mostly insignificant compared to the pre-test scores.

Visitors who viewed the exhibit in the "closed room" test who were paced at a 40-second-interval per exhibit showed significantly more learning (60% correct answers) than visitors paced at the "normal" rate of 15-seconds per exhibit, but learning was the same for visitors who stayed as long as they wanted (the "saturation" condition). The hoped-for mastery level was not achieved. Viewing the exhibit in the proper sequence (to encounter the "narrative thread" in the right order) had no significant impact on learning.

Visitors who were given a 20-minute lecture tour of the exhibit scored significantly higher, 70%.

Problems with the exhibit room design (a busy passageway), lack of a clear starting point, uniformly designed cases (providing little variety), complex labels (although the *intention* had been to keep them simple) and difficult test items on the questionnaire all could have contributed to the apparently low level of learning achieved by visitors to the Magnetism Room.

Suggestions were made for improvements in design, architecture, motivational devices and clarity and simplicity of the exhibit messages to match the level of educational intention on the part of the audience, as well as providing supplementary exhibit techniques (e.g., demonstrations and lectures). The *potential* of museum exhibits to teach science to a general audience was still strongly assumed, although not well demonstrated, in this experimental gallery.

Science on display: A study of the U.S. science exhibit—Seattle World's Fair, 1962. J.B. TAYLOR, 1963. Seattle: University of Washington, Institute for Sociological Research.

Statement of the problem/main purpose of the study:

This 184-page, detailed, descriptive report considers the five halls of the United States science pavilion at the world's fair as an elaborate attempt at mass education and documents its success in communicating information to visitors and changing their attitudes about science. At the time, only a few studies had attempted to research the effectiveness of exhibits, and several groups (e.g., NSF, AAAS) were interested in this general, large-scale investigation.

In addition to looking for evidence of attitude changes and knowledge gains, those changes were correlated with the kinds of displays and the techniques used. General demographics of the fair-goers were also collected and compared.

Sample selected for the study:

More than 9,000 adult visitors to the fair were randomly selected to be part of this study. The original intention was to sample from visitors entering and exiting the entire pavilion and compare them to visitors entering the fair, but several logistical problems precluded that plan, including fair-goers refusals to participate in the study (as they entered the park) and unplanned and irregular traffic flow through the pavilion (e.g., entering through exits, not going completely through the sequence of Halls I to V).

Overview of methods used in the study/procedures:

Due to the logistic problems mentioned above, the focus of the study of learning was concentrated on one hall, Hall IV, "The Methods of Science."

Interviews using multiple choice, rating scales, and semantic differentials were the primary method used to gather information. Tools to aid the interviews included magnetic questionnaire boards, and an automated teaching machine, which proved unreliable and was replaced with face-to-face methods.

Time-lapse cameras recorded crowd flow, and crowd counts were done by student observers hired during two sampling periods.

A smaller interview study was conducted with visitors exiting the pavilion to gather general reactions using free-response questions.

The report contain considerable interesting details about sampling procedures, test question formulation and analysis methods.

Outcomes/major findings:

Many of the conclusions in this report were considered tentative or were stated as propositions, rather than conclusions because 1) the sampling problems caused samples to be less random than was desired (e.g., 70% refusal rate at the entrance gate), and 2) not all of the analysis had been carried out at the time of the report. (Further analysis was alluded to be in an upcoming doctoral thesis, but no such thesis author—A. Dorius—was found by the writing of this abstract.)

Types of visitors attracted to the pavilion

The majority came from Washington, California and Oregon; were more highly educated, with higher income than average. More men than women attended.

Attitudes about science and scientists

Fair-goers held scientists and science in high regard, but had some reservations, and were somewhat vague about their understanding of the process of science. Changes in attitude after viewing exhibits were significant but of slight magnitude, the majority of changes occurring in response to a film in Hall I, after which visitors thought scientists were more eccentric and academic; they saw science as warmer and more feminine, but their general concept of science was more vague. These changes seemed to "recover" after visitors had been exposed to Hall II. Halls III and IV did not have any patterns of changed attitudes, and there was no evidence that visitors overall understanding of science had improved. None of the expected attitude changes had actually appeared.

Learning from exhibits

Small but significant increases in the percentage of correct answers, indicating information retention, occurred more with one of the three "landmark" exhibits: the Satellite Tracking Station, the Biological Laboratory, and the Behavior section, all on the path of the main crowd flow. Percent of correct answers pre-visit were 39% to 61%; post-viewing scores were 43% to 61% (ranges for all items). Exhibits chosen as being "enjoyed" were also the ones for which the greatest information retention occurred. Low retention was partly attributed to the "holiday mood" of visitors and their lack of serious and prolonged study of exhibits.

Effective exhibit techniques

Exhibits with movement and sound tended to be associated with learning gains. "Live" exhibits were the most visited and the most popular. The single most popular exhibit in Hall IV was the monkeys.

Crowd flow

About 67% of the people on the fairgrounds visited the pavilion. Thirty-seven to 54% of the people did not begin in Hall I.

In Hall IV, analysis of time lapse photography suggested that four variables influenced crowd flow throughout the space: three of them had to do with the heavy pulses of crowding caused by people leaving films and demonstrations; the other concerned the extreme popularity of the animal exhibits in the Behavior section. Designers of the pavilion apparently did not take into account the overwhelming influence that crowding would have on people's behavior in the Halls.

Overall impressions and reactions

The minority of respondents to a comparatively small survey of visitors (n = 114) as they exited the Science Exhibit complex said that they had seen the total pavilion (i.e., all five halls), and there was no specific route taken by the majority of visitors. Typically, little time was spent, considering the size and complexity of the pavilion. Reactions were favorable, yet vague, and few people saw Hall IV ("The Methods of Science") as communicating anything about the methods of science.

Favorable comments tended to center on the films and the architecture. Negative comments (made by 52% of the visitors, the rest had no complaints) included the difficulty of understanding the displays and the crowded conditions of viewing.

The final chapter of this report contains thoughts about the psychology of exhibit design. Suggestions are made that exhibit planning and design should consider the process of viewing exhibits—the total experience—and the interactions of exhibits in a sequence, not a static view. Taking account of the time dimension required of each exhibit experience is important.

Four psychological principles with implications for exhibit design are offered:

1. Perception starts with a general gestalt, then goes to finer discriminations.

2. Discriminations are aided by contrast.

3. The amount of the observer's activity in forming a percept will influence its impact.

4. People feel pleased when they are able to form a gestalt, and have a sense of dissatisfaction when they can't.